Praise for Hope Jahren's

LAB GIRL

Winner of the American Association for the Advancement of Science/Subaru *Science Books & Films* Prize for Excellence in Science Books

A *Kirkus Reviews* Best Memoir of the Year

"Spirited. . . . Stunning. . . . Moving."

—*The New York Times Book Review*

"*Lab Girl* made me look at trees differently. It compelled me to ponder the astonishing grace and gumption of a seed. Perhaps most important, it introduced me to a deeply inspiring woman—a scientist so passionate about her work I felt myself vividly with her on every page. This is a smart, enthralling, and winning debut."

—Cheryl Strayed

"[A] powerful new memoir. . . . Jahren is a remarkable scientist who turns out to be a remarkable writer as well. . . . Think Stephen Jay Gould or Oliver Sacks. But Hope Jahren is a woman in science, who speaks plainly to just how rugged that can be. And to the incredible machinery of life around us." —*On Point*, NPR

"Lyrical . . . illuminating. . . . Offers a lively glimpse into a scientifically inclined mind." —*The Wall Street Journal*

"Some people are great writers, while other people live lives of adventure and importance. Almost no one does both. Hope Jahren does both. She makes me wish I'd been a scientist."

—Ann Patchett, author of *State of Wonder*

"*Lab Girl* surprised, delighted, and moved me. I was drawn in from the start by the clarity and beauty of Jahren's prose. . . . With *Lab Girl*, Jahren joins those talented scientists who are able to reveal to us the miracle of this world in which we live."

—Abraham Verghese, author of *Cutting for Stone*

"Revelatory. . . . A veritable jungle of ideas and sensations." —*Slate*

"Warm, witty. . . . Fascinating. . . . Jahren's singular gift is her ability to convey the everyday wonder of her work: exploring the strange, beautiful universe of living things that endure and evolve and bloom all around us, if we bother to look." —*Entertainment Weekly*

"Deeply affecting. . . . A totally original work, both fierce and uplifting. . . . A belletrist in the mold of Oliver Sacks, she is terrific at showing just how science is done. . . . She's an acute observer, prickly, and funny as hell." —*Elle*

"Magnificent. . . . [A] gorgeous book of life. . . . Jahren contains multitudes. Her book is love as life. Trees as truth." —*Chicago Tribune*

"Mesmerizing. . . . Deft and flecked with humor . . . a scientist's memoir of a quirky, gritty, fascinating life. . . . Like Robert Sapolsky's *A Primate's Memoir* or Helen Macdonald's *H Is for Hawk*, it delivers the zing of a beautiful mind in nature." —*The Seattle Times*

"Jahren's memoir [is] the beginning of a career along the lines of Annie Dillard or Diane Ackerman." —*Minneapolis Star Tribune*

"A scientific memoir that's beautifully human." —*Popular Science*

"Breathtakingly honest. . . . Gorgeous. . . . At its core, *Lab Girl* is a book about seeing—with the eyes, but also the hands and the heart."

—*American Scientist*

Hope Jahren

LAB GIRL

Hope Jahren is an award-winning scientist who has been pursuing independent research in paleobiology since 1996, when she completed her Ph.D. at the University of California, Berkeley and began teaching and researching, first at the Georgia Institute of Technology and then at Johns Hopkins University. She is the recipient of three Fulbright Awards and is one of four scientists, and the only woman, to have been awarded both of the Young Investigator Medals given in the earth sciences. She was a tenured professor at the University of Hawai'i in Honolulu from 2008 to 2016, where she built the Isotope Geobiology Laboratories, with support from the National Science Foundation, the U.S. Department of Energy, and the National Institutes of Health. She currently holds the J. Tuzo Wilson Professorship at the University of Oslo, Norway.

hopejahrensurecanwrite.com

jahrenlab.com

LAB GIRL

LAB GIRL

Hope Jahren

VINTAGE BOOKS
A Division of Penguin Random House LLC
New York

FIRST VINTAGE BOOKS EDITION, MARCH 2017

Lab Girl is a work of nonfiction. To protect the privacy of others, certain names have been changed, characters conflated, and some incidents condensed.

Portions of chapter 1 originally appeared in *Nautilus*, Issue 34: Adaptation (nautil.us) as "My Family, My Science" on March 3, 2016.

The Library of Congress has cataloged the Knopf edition as follows:
Names: Jahren, Hope.
Title: Lab girl / Hope Jahren.
Description: First edition. | New York : Alfred A. Knopf, 2016. | Includes bibliographical references.
Identifiers: LCCN 2015024305 | Subjects: LCSH: Jahren, Hope. | Biologists—United States—Biography. | Geobiology—Research—Anecdotes.
Classification: LCC QH31.J344 A3 2016 | DDC 570.92—dc23
LC record available at http://lccn.loc.gov/2015024305

Vintage Books Trade Paperback ISBN: 978-1-101-87372-4
eBook ISBN: 978-1-101-87494-3

Book design by M. Kristen Bearse

www.vintagebooks.com

Printed in the United States of America
10

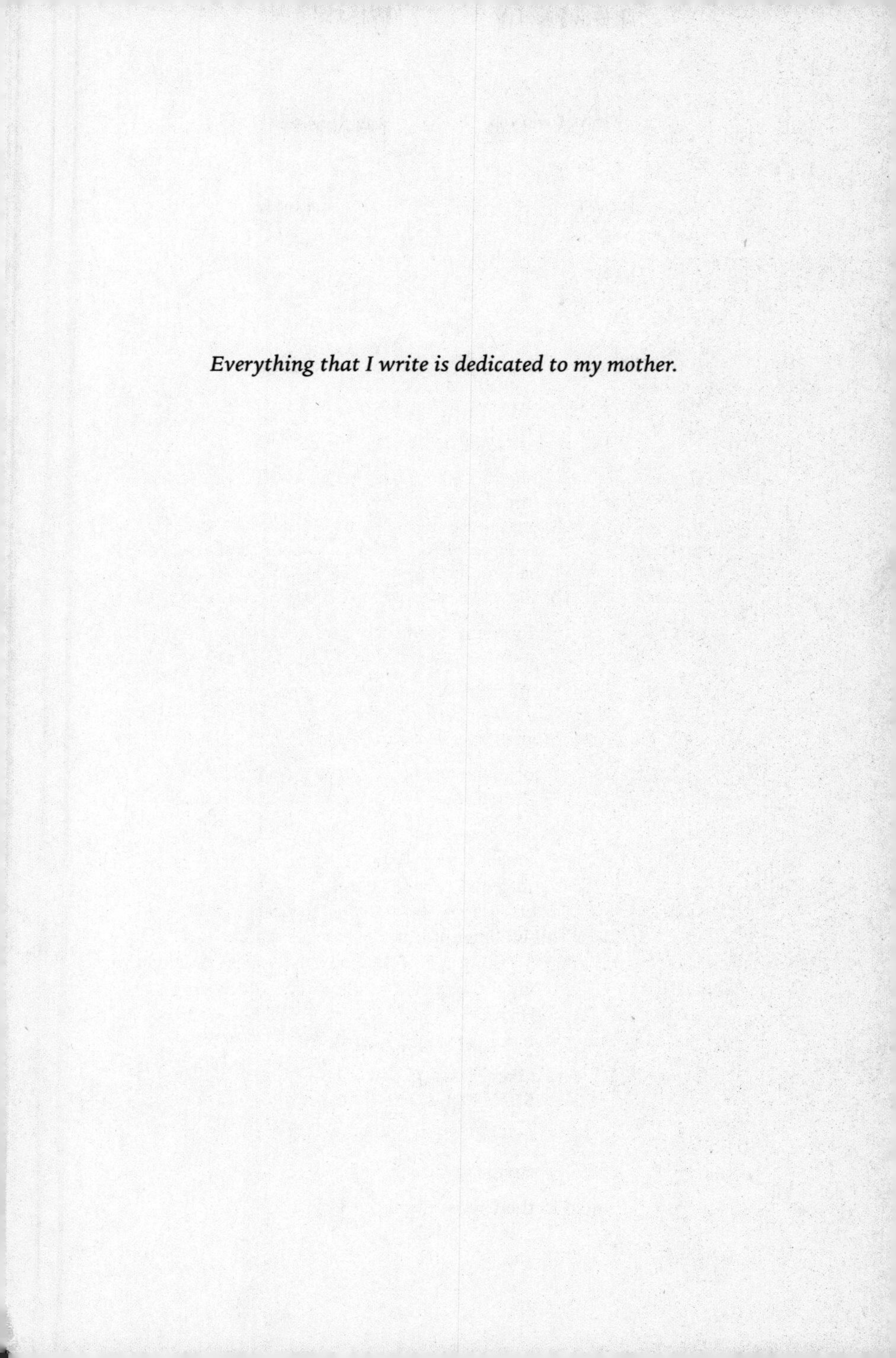

Everything that I write is dedicated to my mother.

The more I handled things and learned their names and uses, the more joyous and confident grew my sense of kinship with the rest of the world.

—Helen Keller

Contents

LAB GIRL

Prologue

PEOPLE LOVE THE OCEAN. People are always asking me why I don't study the ocean, because, after all, I live in Hawaii. I tell them that it's because the ocean is a lonely, empty place. There is six hundred times more life on land than there is in the ocean, and this fact mostly comes down to plants. The average ocean plant is one cell that lives for about twenty days. The average land plant is a two-ton tree that lives for more than one hundred years. The mass ratio of plants to animals in the ocean is close to four, while the ratio on land is closer to a thousand. Plant numbers are staggering: there are eighty billion trees just within the protected forests of the western United States. The ratio of trees to people in America is well over two hundred. As a rule, people live among plants but they don't really see them. Since I've discovered these numbers, I can see little else.

So humor me for a minute, and look out your window.

What did you see? You probably saw things that people make. These include other people, cars, buildings, and sidewalks. After just a few years of design, engineering, mining, forging, digging, welding, bricklaying, window-framing, spackling, plumbing, wiring, and painting, people can make a hundred-story skyscraper capable of casting a thousand-foot shadow. It's really impressive.

Now look again.

Did you see something green? If you did, you saw one of the few things left in the world that people cannot make. What you saw was invented more than four hundred million years ago near the equator. Perhaps you were lucky enough to see a tree. That tree was designed about three hundred million years ago. The mining of the atmosphere, the cell-laying, the wax-spackling, plumbing, and pigmentation took a few months at most, giving rise to nothing more or less perfect than a leaf. There are about as many leaves on one tree as there are hairs on your head. It's really impressive.

Now focus your gaze on just one leaf.

People don't know how to make a leaf, but they know how to destroy one. In the last ten years, we've cut down more than fifty billion trees. One-third of the Earth's land used to be covered in forest. Every ten years, we cut down about 1 percent of this total forest, never to be regrown. That represents a land area about the size of France. One France after another, for decades, has been wiped from the globe. That's more than one trillion leaves that are ripped from their source of nourishment every single day. And it seems like nobody cares. But we should care. We should care for the same basic reason that we are always bound to care: because someone died who didn't have to.

Someone died?

Maybe I can convince you. I look at an awful lot of leaves. I look at them and I ask questions. I start by looking at the color: Exactly what shade of green? Top different from the bottom? Center different from the edges? And what about the edges? Smooth? Toothed? How hydrated is the leaf? Limp? Wrinkled? Flush? What is the angle between the leaf and the stem? How big is the leaf? Bigger than my hand? Smaller than my fingernail? Edible? Toxic? How much sun does it get? How often does the rain hit it? Sick? Healthy? Important? Irrelevant? Alive? Why?

Now *you* ask a question about *your* leaf.

Guess what? You are now a scientist. People will tell you that you have to know math to be a scientist, or physics or chemistry. They're wrong. That's like saying you have to know how to knit to be a housewife, or that you have to know Latin to study the Bible. Sure, it helps, but there will be time for that. What comes first is a question, and you're already there. It's not nearly as involved as people make it out to be.

So let me tell you some stories, one scientist to another.

Part One

ROOTS AND LEAVES

1

THERE IS NOTHING in the world more perfect than a slide rule. Its burnished aluminum feels cool against your lips, and if you hold it level to the light you can see God's most perfect right angle in each of its corners. When you tip it sideways, it gracefully transfigures into an extravagant rapier that is also retractable with great stealth. Even a very little girl can wield a slide rule, the cursor serving as a haft. My memory cannot separate this play from the earliest stories told to me, and so in my mind I will always picture an agonized Abraham just about to almost sacrifice helpless little Isaac with his raised and terrible slide rule.

I grew up in my father's laboratory and played beneath the chemical benches until I was tall enough to play on them. My father taught forty-two consecutive years' worth of introductory physics and earth science in that laboratory, nestled within a community college deep in rural Minnesota; he loved his lab, and it was a place that my brothers and I loved also.

The walls were made of cinder blocks slathered in thick cream-colored semigloss paint, but you could feel the texture of the cement underneath if you closed your eyes and concentrated. I remember deciding that the black rubber wainscoting must have been attached with adhesive, because I couldn't find any nail holes anywhere when I measured its whole length with the yellow surveying tape that extended to a full thirty meters. There were long workbenches where five college boys were to sit side by side, all facing the same direction. These black countertops felt cool as a tombstone and were made of something just as timeless, something that acid couldn't burn and a hammer couldn't smash (but don't try). The benches were strong enough for you to stand on the edge of and couldn't be scratched even with a rock (but don't try).

Evenly spaced across the benches were braces of impossibly shiny silver nozzles with handles that took all your strength to turn ninety degrees, and when you did the one that said "gas" did nothing because it wasn't hooked up, but the one that said "air" blew with such an exhilarating rush that you kind of wanted to put your mouth on it (but don't try). The whole place was clean and open and empty, but each drawer contained a fascinating array of magnets, wire, glass, and metal that were all useful for something; you just had to figure out what it was. In the cupboard by the door there was pH testing tape, which was like a magic trick only better because instead of just showing a mystery it also solved one: you could see the difference in color and thus pH between a drop of spit and a drop of water or root beer or urine in the bathroom but not blood because you can't see through it (so don't try). These were not kids' toys; they were serious things for grown-ups, but you were a special kid because your dad had that huge ring of keys, so you could play with the equipment anytime you went there with him, because he never, ever said no when you asked him to take it all out.

In my memory of those dark winter nights, my father and I own the whole science building, and we walk about like a duke and his sovereign prince, too preoccupied in our castle to bother about our frozen duchy. As my father prepared for class the next day, I would work backward through each canned experiment and demonstration, making sure the college boys would have the easy success toward which they were predisposed. We pored over the equipment and fixed what was broken, and my father taught me how to preemptively take things apart and study how they work, so that as they inevitably failed I'd be able to restore them. He taught me that there is no shame in breaking something, only in not being able to fix it.

At eight o'clock we would start our walk home, so I could be in bed by nine. First we'd stop by my father's tiny, windowless office, which was bare of decoration except for the pencil holder that I had made for him out of clay. From there we'd collect our coats, hats, scarves, and the other things that my mother had knitted for me because she had never had decent ones when she was a little girl. As I wrestled my sturdy boots over an extra pair of socks, the smell of warm, wet wool mixed with that of wood shavings as my father sharpened each of the pencils that we had dulled. He would then briskly button up his big coat and don his deerskin mittens and tell me to check that my hat was fully covering both of my ears.

Always the last to leave the building for the day, he would walk the halls twice, first to confirm that all the doors to the outside were locked, and then to turn off the lights, one by one, as I trotted along behind him, fleeing the pursuing darkness. Finally, at the back entrance, my father would let me reach up and swipe off the last set of light switches, and we'd walk outside. He'd pull the door shut behind us and then check it twice to make sure that the lock had set.

Thus sealed out into the cold, we'd stand on the loading dock and look up at the frozen sky and into the terminal coldness of space and see light that had been emitted years ago from unimaginably hot fires that were still burning on the other side of the galaxy. I didn't know any of the constellations that people used to name the stars above me, and I never asked what they might be, though I am certain that my father knew each one and the story behind it. We had long since established the habit of not speaking as we walked the two miles home; silent togetherness is what Scandinavian families do naturally, and it may be what they do best.

The community college where my father worked was situated at the western end of our little hometown, the incorporated portion of which spanned four miles from truck stop to truck stop. My three older brothers and I lived with our parents in a big brick house located south of Main Street, four blocks west of where my father had grown up in the 1920s, eight blocks east of where my mother had grown up in the 1930s, one hundred miles south of Minneapolis, and five miles north of the Iowa border.

Our path through town took us past the clinic where the same doctor who had delivered me occasionally swabbed my throat to test for strep infections, past the toothpaste-blue water tower that constituted the tallest structure in town, past the high school that was manned by teachers who had once been my father's students. When we passed under the eaves troughs of the Presbyterian church where my father and mother had their first date at a Sunday School picnic in 1949, were married in 1953, had me baptized in 1969, and where our family spent every Sunday morning without exception, my father would lift me up so that I could break off a thick icicle. I would kick it along like a hockey puck while we walked, and it would ring out every ten steps or so as it ricocheted off the sides of the hard-packed snowbanks.

We made our way down hand-shoveled sidewalks, past thickly insulated houses that sheltered families who were no doubt partaking of silences similar to our own. In almost every one of these houses lived someone that we knew. From playpen to prom, I grew up with the sons and daughters of the girls and boys whom my mother and father had played with when they were children, and none of us could remember a time when we all hadn't known each other, even if our deeply bred reticence kept us from knowing much about each other. It wasn't until I was seventeen and moved away to college that I discovered how the world is mostly populated by strangers.

When I heard a weary monster sighing on the other side of town, I understood that it was twenty-three minutes after eight o'clock and the train was pulling out of the factory, as it did every night. I heard the great iron brakes wrench and then relax as a string of empty tank cars started to drag northward, toward Saint Paul, where they would each be filled with thirty thousand gallons of brine. In the morning we would hear the train return, and the exhausted monster would again sigh as its burden was pumped into the bottomless reservoir of salt made necessary by the factory's continuous manufacture of bacon.

The train tracks ran north-south, isolating one corner of my little town, upon which still stands what is perhaps the most magnificent slaughterhouse of the Midwest. Starting at its killing chute, upwards of twenty thousand animals are processed for their meat every single day.

Mine was one of the few families I knew of that was not directly employed by the factory, but our extended lineage had worked there plenty. My great-grandparents, like practically everybody else's in that town, had come to Minnesota as part of a mass emigration from Norway that began in about 1880. And like everybody else in my hometown, this is pretty much all that I knew about my ancestors. I suspected that they hadn't relocated to the coldest place on Earth and then taken up disemboweling pigs because things were going well in Europe, but it had never occurred to me to ask for the story.

I never met my grandmothers—they had both died before I was born. I could remember my grandfathers, who had died when I was four and seven, respectively, but I couldn't remember any occasion when either of them had spoken directly to me. My father had been an only child, but I think my mother had more than ten siblings, many of whom I never met. Entire years passed between our visits with my aunts and uncles, even though some of them lived in the same small town that we did. I didn't much notice as my three older brothers grew up and left our home one by one, as it was not unusual for us to go days without finding anything to say to each other.

The vast emotional distances between the individual members of a Scandinavian family are forged early and reinforced daily. Can you imagine growing up in a culture where you can never ask anyone anything about themselves? Where "How are you?" is considered a personal question that one is not obligated to answer? Where you are trained to always wait for others to first mention what is troubling them, even as you are trained to never mention what is troubling you? It must be a survival skill left over from the old Viking days, when long silences were required to prevent unnecessary homicides during the long, dark winters when quarters were close and supplies were dwindling.

While I was a child, I assumed that the whole world acted like we did, and so it confused me when I moved out of state and met people who effortlessly gave each other the simple warmth and casual affection that I had craved for so long. I then had to learn to live in a world where when people don't talk to each other, it is because they don't know each other, not because they do.

By the time my father and I had crossed Fourth Street (or "Kenwood Avenue," as he called it; he had learned the streets as a child in the 1920s, long before they were numbered, and never adopted the new system), we could see the front door of our big brick house. It was the house that my mother had dreamed of living in as a child, and after my parents were married they had saved for eighteen years in order to buy it. Despite my having walked briskly—it was always an effort to keep up with my father—my fingers were chilled such that I knew it would be painful when they warmed up. Once it gets to a certain point below zero, the thickest mittens in the world won't keep your hands warm, and I was glad that the walk was almost over. My father turned the heavy iron handle, pushed with his shoulder, and opened our oaken front door. We went inside the house, into a different kind of cold.

In the foyer, I sat down and wrestled off my boots, then began to molt coats and sweaters. My father hung our clothes in the heated closet, and I knew that they would be waiting for me, warm and dry, when it was time to walk to school the following morning. I could hear my mother in the kitchen unloading the dishwasher, the butter knives clanging together as she dropped them into the silverware drawer and then slammed it shut. She was always angry and I could never piece together why. With the self-focus peculiar to children, I convinced myself that it must be because of something that I had said or done. In the future, I vowed to myself, I would guard my words better.

I went upstairs, changed into my flannel pajamas, and put myself to bed. My bedroom faced south toward the frozen pond where I would spend all day Saturday ice-skating—if it had warmed up enough by then. The wool carpet was dusky-blue and the walls had been papered in complementary damask. The room had originally been designed for twin girls, with two built-in desks, two built-in vanities, and so on. On the nights when I couldn't sleep, I would sit at my window seat and trace the feathery ice crystals across the glass with my finger, trying not to look at the vacant seat in front of the other window where a sister should have been.

The fact that I remember so much cold and darkness from my childhood isn't surprising, given the fact that I grew up in a place where there was snow on the ground for nine months out of each year.

Descending into and then surfacing from winter formed the driving rhythm of our lives, and as a child I assumed that people everywhere watched as their summer world died, confident in its eventual resurrection, having been tested so often within a crucible of ice.

Every year I saw the first stuttering flakes of September crescendo into the spilling white heaps of December, then petrify into the deep, icy emptiness of late February, eventually to be varnished as a grand, frictionless expanse by the stinging April sleet. Our Halloween costumes as well as our Easter dresses were sewn such that they could be worn inside our snowsuits, and Christmastime wrapped us in wool, velvet, and more wool. The one summertime activity that I remember vividly is working in the garden with my mother.

In Minnesota, the spring thaw happens all at once when the frozen ground yields to the sun in one day, wetting the spongy soil from within. On the first day of spring, you can reach into the ground and easily pull up great, loose clumps of dirt as if they were handfuls of too-fresh devil's food cake and watch the fat pink earthworms come writhing out and fling themselves joyfully back into the hole. There is not even a hint of clay within the soils of southern Minnesota; they have lain like a rich black blanket over the limestone of the region for a hundred thousand years, periodically chased off by glaciers. They are richer than any prefertilized potting soil that you can buy at the hardware store; anything will grow in a Minnesota garden, and there's no need to water or fertilize—the rain and the worms will supply everything that is needed—but the growing season is short, so there's no time to be wasted.

My mother wanted two things from her garden: efficiency and productivity. She favored sturdy, independent vegetables like Swiss chard and rhubarb, the ones that could be relied upon to yield in abundance and seemed only to thrive in response to frequent harvesting. She had neither the time nor the sympathy required to nurse lettuce or prune tomatoes; instead she preferred the radishes and carrots that could tend to their own needs quietly underground. Even the flowers that she grew were selected for their toughness: the golf ball–sized buds of peonies that spilled out petals as they swelled into pink blossoms the size of cabbages, the leathery tiger lilies, and the fat, bearded irises that barged out of their bulbs without fail, spring after spring after spring.

Each May Day, my mother and I poked individual seeds into the ground, and then a week later we plucked out the ones that hadn't grown, replacing them and immediately starting over. By the end of June the whole crop was well on its way and the world around us was so green that it seemed impossible that it could ever have been otherwise. By July, the sweating leaves of all of these plants had pumped the air so full of moisture that the humidity caused the electrical lines to buzz and crackle overhead.

My strongest memory of our garden is not how it smelled, or even looked, but how it sounded. It might strike you as fantastic, but you really can hear plants growing in the Midwest. At its peak, sweet corn grows a whole inch every single day and as the layers of husk shift slightly to accommodate this expansion, you can hear it as a low continuous rustle if you stand inside the rows of a cornfield on a perfectly still August day. As we dug in our garden, I listened to the lazy buzzing of bees as they staggered drunkenly from flower to flower, the petty, sniping chirps of the cardinals remarking upon our bird feeder, the scraping of our trowels through the dirt, and the authoritative whistle of the factory, blown each day at noon.

My mother believed that there was a right way and a wrong way to do everything, and that doing it wrong meant doing it over, preferably a few times. She knew how to stitch a different tension into each of the buttons on a shirt, based on how often it would be called into use. She knew the best way to pick elderberries on a Monday such that their stems wouldn't clog the old tin colander on Wednesday, when we strained them after stewing them all day Tuesday. Thinking two steps ahead in every conceivable direction, she never doubted herself, and I figured that there was nothing in the world she didn't know how to do.

In fact, my mother did know how to do—and had kept doing—a lot of things that weren't strictly necessary now that the Great Depression was done and war shortages weren't in effect and President Ford had assured us that all of our nightmares were over. She saw her own rags-to-relative-riches life story as a hard-fought victory over something evil, decided that her children should keep fighting in order to deserve its legacy, and proceeded to toughen us for a struggle that never came.

Whenever I looked at my mother, it was difficult for me to believe that the well-spoken and smartly dressed woman before me could ever have been a dirty, hungry, and scared child. Only her hands gave her away: they were far too durable for the life she now led, and I sensed that she could have grabbed the rabbit that plagued our garden and wrung its neck without thinking, had it been stupid enough to get too close to her.

When you grow up around people who don't speak very much, what they do say to you is indelible. As a child, my mother had been both the poorest and the smartest girl in Mower County. During her senior year of high school, she was awarded an honorable mention in the ninth annual nationwide Westinghouse Science Talent Search. This was an unusual recognition for a female growing up in a rural area, and although it counted only as a near miss for the real prize, it put her in good company. Other 1950 also-rans included Sheldon Glashow, who went on to win a Nobel Prize in Physics, and Paul Cohen, who won the Fields Medal in 1966—the highest honor given in the field of mathematics.

Unfortunately for my mother, scoring an honorable mention carried with it a one-year honorary junior membership to the Minnesota Academy of Science, not the college scholarship that she had been hoping for. Undeterred, she moved to Minneapolis anyway and tried to support herself while studying chemistry at the University of Minnesota, but soon came to the realization that she couldn't attend the long afternoon laboratory sections and still put in enough babysitting hours to pay her tuition. In 1951, the university experience was designed for men, usually men with money, or at the very least men who had job options outside of being some family's live-in nanny. She moved back to our hometown, married my father, gave birth to four children, and threw herself into twenty years of raising them. Determined to earn her bachelor's degree once the last of her children was at least in preschool, she re-enrolled in the University of Minnesota. Her options were limited to correspondence courses, so she chose English literature. Because I spent my days mostly in her care, it became natural for her to include me in her studies.

We plowed through Chaucer, and I learned to assist her using the Middle English dictionary. One year we spent the winter painstakingly noting each instance of symbolism within *Pilgrim's Progress* on separate recipe cards, and I was delighted to see our pile grow to be thicker than the book itself. She set her hair in curlers while listening to records of Carl Sandburg's poems over and over, and instructed me on how to hear the words differently each time. After discovering Susan Sontag, she explained to me that even meaning itself is a constructed concept, and I learned how to nod and pretend to understand.

My mother taught me that reading is a kind of work, and that every paragraph merits exertion, and in this way, I learned how to absorb difficult books. Soon after I went to kindergarten, however, I learned that reading difficult books also brings trouble. I was punished for reading ahead of the class, for being unwilling to speak and act "nicely." I didn't know why I simultaneously feared and adored my female teachers, but I did know that I needed their attention, positive or negative, at all times. Tiny but determined, I navigated the confusing and unstable path of being what you are while knowing that it's more than people want to see.

Back at home, while my mother and I gardened and read together, I vaguely sensed that there was something we weren't doing, something affectionate that normal mothers and daughters naturally do, but I couldn't figure out what it was, and I suppose she couldn't either. We probably do love each other, each in our own stubborn way, but I'm not entirely sure, probably because we have never openly talked about it. Being mother and daughter has always felt like an experiment that we just can't get right.

When I was five I came to understand that I was not a boy. I still wasn't sure what I was, but it became clear that whatever I was, it was less than a boy. I saw that my brothers, who were five, ten, and fifteen years older than I, could do all of our laboratory play in the outside world. In Cub Scouts they raced model cars and built and set off rockets. In shop class they used tools big and powerful enough to be mounted on the wall or suspended from the ceiling. When we watched Carl Sagan and Mr. Spock and Doctor Who and the Professor, we never even commented on Nurse Chapel or Mary Ann in the background. I retreated further into my father's laboratory, as the place where I could most freely explore the mechanical world.

It made sense, in a way. I was the one who was like our father, or at least I thought so. The differences between us were purely cosmetic: my father looked just like a scientist was supposed to. He was tall and pale and clean-shaven, thin-almost-gaunt in his khakis and white shirt and horn-rimmed glasses, complete with a pronounced Adam's apple. When I was five I also decided that the real me looked exactly like that, even though on the outside I was disguised as a girl.

While pretending to be a girl I spent my time deftly grooming myself and gossiping with my girlfriends about who liked whom and what if they didn't. I could jump rope for hours and sew my own clothes and make anything anybody wanted to eat from scratch in three different ways. But in the late evenings I would accompany my dad to his laboratory, when the building was empty but well lit. There I transformed from a girl into a scientist, just like Peter Parker becoming Spider-Man, only kind of backward.

As much as I desperately wanted to be like my father, I knew that I was meant to be an extension of my indestructible mother: a do-over to make real the life that she deserved and should have had. I left high school a year early to take a scholarship at the University of Minnesota—the same school that my mother, my father, and all of my brothers had attended.

I started out studying literature, but soon discovered that science was where I actually belonged. The contrast made it all the clearer: in science classes we did things instead of just sitting around talking about things. We worked with our hands and there were concrete and almost daily payoffs. Our laboratory experiments were predesigned to work perfectly and elegantly every time, and the more of them that you did, the bigger the machines and the more exotic were the chemicals that they let you use.

Science lectures dealt with social problems that still could be solved, not defunct political systems for which both the proponents and the opponents had died before my birth. Science didn't talk about books that had been written to analyze other books that had originally been written as retellings of ancient books; it talked about what was happening now and of a future that might yet be. The very attributes that rendered me a nuisance to all of my previous teachers—my inability to let things go coupled with my tendency to overdo everything—were exactly what my science professors liked to see. They accepted me despite the fact that I was just a girl, and assured me of what I already suspected: that my true potential had more to do with my willingness to struggle than with my past and present circumstances. Once again I was safe in my father's laboratory, allowed to play with all of the toys for as long as I wanted.

People are like plants: they grow toward the light. I chose science because science gave me what I needed—a home as defined in the most literal sense: a safe place to be.

Growing up is a long and painful process for everyone, and the only thing I ever knew for certain was that someday I would have my own laboratory because my father had one. In our tiny town, my father wasn't *a* scientist, he was *the* scientist, and being a scientist wasn't his job, it was his identity. My desire to become a scientist was founded upon a deep instinct and nothing more; I never heard a single story about a living female scientist, never met one or even saw one on television.

As a female scientist I am still unusual, but in my heart I was never anything else. Over the years I have built three laboratories from scratch, given warmth and life to three empty rooms, each one bigger and better than the last. My current laboratory is almost perfect—located in balmy Honolulu and housed within a magnificent building that is frequently crowned by rainbows and surrounded by hibiscus flowers in constant bloom—but somehow I know that I will never stop building and wanting more. My laboratory is not "room T309," as stated on my university's blueprints; it is "the Jahren Laboratory" and it always will be, no matter where it is located. It bears my name because it is my home.

My laboratory is a place where the lights are always on. My laboratory has no windows, but it needs none. It is self-contained. It is its own world. My lab is both private and familiar, populated by a small number of people who know one another well. My lab is the place where I put my brain out on my fingers and I do things. My lab is a place where I move. I stand, walk, sit, fetch, carry, climb, and crawl. My lab is a place where it's just as well that I can't sleep, because there are so many things to do in the world besides that. My lab is a place where it matters if I get hurt. There are warnings and rules designed to protect me. I wear gloves, glasses, and closed-toed shoes to shield myself against disastrous mistakes. In my lab, whatever I need is greatly outbalanced by what I have. The drawers are packed full with items that might come in handy. Every object in my lab—no matter how small or misshapen—exists for a reason, even if its purpose has not yet been found.

My lab is a place where my guilt over what I haven't done is supplanted by all of the things that I am getting done. My uncalled parents, unpaid credit cards, unwashed dishes, and unshaved legs pale in comparison to the noble breakthrough under pursuit. My lab is a place where I can be the child that I still am. It is the place where I play with my best friend. I can laugh in my lab and be ridiculous. I can work all night to analyze a rock that's a hundred million years old, because I need to know what it's made of before morning. All the baffling things that arrived unwelcome with adulthood—tax returns and car insurance and Pap smears—none of them matter when I am in the lab. There is no phone and so it doesn't hurt when someone doesn't call me. The door is locked and I know everyone who has a key. Because the outside world cannot come into the lab, the lab has become the place where I can be the real me.

My laboratory is like a church because it is where I figure out what I believe. The machines drone a gathering hymn as I enter. I know whom I'll probably see, and I know how they'll probably act. I know there'll be silence; I know there'll be music, a time to greet my friends, and a time to leave others to their contemplation. There are rituals that I follow, some I understand and some I don't. Elevated to my best self, I strive to do each task correctly. My lab is a place to go on sacred days, as is a church. On holidays, when the rest of the world is closed, my lab is open. My lab is a refuge and an asylum. It is my retreat from the professional battlefield; it is the place where I coolly examine my wounds and repair my armor. And, just like church, because I grew up in it, it is not something from which I can ever really walk away.

My laboratory is a place where I write. I have become proficient at producing a rare species of prose capable of distilling ten years of work by five people into six published pages, written in a language that very few people can read and that no one ever speaks. This writing relates the details of my work with the precision of a laser scalpel, but its streamlined beauty is a type of artifice, a size-zero mannequin designed to showcase the glory of a dress that would be much less perfect on any real person. My papers do not display the footnotes that they have earned, the table of data that required painstaking months to redo when a graduate student quit, sneering on her way out that she didn't want a life like mine. The paragraph that took five hours to write while riding on a plane, stunned with grief, flying to a funeral that I couldn't believe was happening. The early draft that my toddler covered in crayon and applesauce while it was still warm from the printer.

Although my publications contain meticulous details of the plants that did grow, the runs that went smoothly, and the data that materialized, they perpetrate a disrespectful amnesia against the entire gardens that rotted in fungus and dismay, the electrical signals that refused to stabilize, and the printer ink cartridges that we secured late at night through nefarious means. I know damn well that if there had been a way to get to success without traveling through disaster someone would have already done it and thus rendered the experiments unnecessary, but there's still no journal where I can tell the story of how my science is done with both the heart and the hands.

Eventually 8:00 a.m. rolls around, the chemicals need to be restocked, the paychecks need to be cut, the plane tickets need to be bought, and so I've lowered my head and written yet another scientific report while the pain, pride, regret, fear, love, and longing have built up deep in my throat unspoken. Working in a lab for twenty years has left me with two stories: the one that I have to write, and the one that I want to.

Science is an institution so singularly convinced of its own value that it cannot bear to throw anything away. This is true even of my father and his slide rules, carefully boxed in the basement of my childhood home and labeled "Standard Linear Slide Rule [25 cm] 30 ct." There are thirty of them, because it is important that each student has his own—scientists do many things, but they do not share equipment. These old slide rules will never be useful again; they've been thoroughly and terminally outmoded, first by calculators, then by desktop computers, and recently by phones. Nobody's name is written on the box, just a label itemizing what's inside. I used to look at it and wish, with an inexplicable yearning, that my father would write my name on the box. But no one owns those slide rules; they just are. And they certainly never belonged to me.

* * *

In 2009, I turned forty years old. By then I had been a professor for fourteen years. It was also the year that we made a significant breakthrough in isotope chemistry, by successfully building a machine that could work side by side with our mass spectrometer.

You probably have a bathroom scale that can tell the difference between a 180-pound man and a 185-pound man. I have a scientific scale that can tell the difference between an atom with twelve nucleons and an atom with thirteen nucleons. Actually, I have two such scales. They are called mass spectrometers, and they are worth about half a million dollars each. The university bought them for me with the not-so-tacit understanding that I would do wonderful and previously impossible things with them and thus further raise the scientific reputation of the institution.

Based on a rough cost-benefit analysis, I need to do about four wonderful and previously impossible things every single year until I fall into the grave in order for the university to break even on me. This is complicated by the fact that the money for every single other thing—chemicals, beakers, Post-it notes, a rag to polish the mass spectrometer—all has to be raised by me through written or verbal supplication for federal or private funding, which is diminishing rapidly on a national level. That is not the most stressful part. The salary of every single person in the lab—aside from my own—also has to be raised by this same mechanism. It would be nice to promise an employee who has sacrificed everything for science and works eighty hours a week more than about six months of job security, but that is not the world within which the research scientist operates. If you're reading this, and you wish to support us, please give me a call. It would be insane of me not to include that sentence.

The year 2009 marked the third year that my team had been working to handcraft an apparatus that could scrub nitrous oxide out of the gases released during the detonation of a homemade explosive. Once we got it working, we were going to attach it to the front end of one of the mass spectrometers and make measurements. We were hoping to contribute a new method of forensic analysis for the chemical aftermath of a terrorist attack, since the number of neutrons in any given substance can serve as a sort of fingerprint. Our idea was to compare, and perhaps link, the chemical fingerprint of post-blast residues with that of the chemical traces gathered from surfaces where the explosives might have been constructed—a kitchen countertop, for example.

We happened to "sell" the idea to the National Science Foundation in 2007—right after the press reported that IEDs (improvised explosive devices) were causing more than half of the deaths of coalition forces in Afghanistan. Not only were we awarded the funding, but the figure had more zeros behind it than I had ever before seen on paper. I wanted to be studying plant growth, but science for war will always pay better than science for knowledge. My devious plan was that we'd put in our forty hours a week on the explosives project and then spend another forty hours moonlighting with our plant biology experiments.

This protocol gave rise to both a splendid exhaustion and increased desperation during the usual setbacks and demi-failures. The chemical reaction that we were tweaking was difficult and recalcitrant: it was easy enough to get the nitrogen out of the explosives residue, but converting the oxygen attached to it proved much trickier than we had assumed, and we had trouble keeping track of the neutrons during the manipulation. In fact, no matter what we analyzed, once we attached it to the mass spectrometer, the readout gave us almost identical values. It was maddening, like asking a human subject to identify a red versus a green light and then having him respond "green" every time, regardless of what you showed him.

At what point do you escort your befuddled subject to the door and begin anew with a different recruit? Well, never, if you are as pigheaded as I am. We had slowed down and become more careful, hoping to exclude the careless imprecisions that a more robust experiment might have tolerated. Soon after that, what we had projected as two-hour tasks in the lab were taking four days to complete, and eight days to complete correctly. We also had to squeeze all of this lab work in between watering, fertilizing, and documenting the growth of a hundred plants every day.

I'll always remember the night that we finally got our explosives analyzer successfully synced up with the mass spectrometer, and it started giving us the standardized values that we knew it should—similar though it was to many other nights of my life. It was a Sunday evening, at the late hour when one first feels Monday begin to threaten. As usual, I was obsessing over our budgets. Because the project was drawing to a close, I could calculate the exact day that the lab would run out of funding. I was sitting in my office poring over chemical prices, casting spells on dimes and trying to alchemize them into dollars, but I still couldn't push back bankruptcy for more than a few months.

The door opened and my lab partner, Bill, came bounding into my office. He plopped himself down in a broken chair and threw some papers onto my desk. "All right, I'm ready to say it. The motherfucker works, and it works beautifully!" he announced.

I began to leaf through his stack of readouts, unsurprised to see that each of the different gas samples now displayed a different, and accurate, value. I am usually ready to pronounce something a success long before Bill is. He always wants to run one more set of standards and do one more calibration before he admits that we've conquered failure.

Bill and I grinned at each other, knowing that we'd pulled it off, yet again. The whole project was a fine example of how we work together: I cook up a pipe dream, embellish it until it is borderline impossible, pitch and sell the idea to a government agency, purchase the supplies, and then dump it all on Bill's desk. From there, Bill produces a first, a second, and then a third prototype, protesting all the while that the idea is an impossible pipe dream. When his fifth design shows promise, and his seventh works (provided you turn it on while wearing a blue shirt and facing east), we are both seduced by the smell of success.

From there, we enter a period of me working days and him working nights, and both of us Tweeting, texting, and Facebooking every single data readout until our homemade creation has proven itself to be as accurate and reliable as my grandmother's Singer treadle sewing machine. Then, after Bill does one more battery of tests—or two, or maybe just also a third one—*then* we are done. It is now my job to revise history for the final report: to narrate the supreme ease with which we've gotten our baby up and running and to itemize what an excellent investment this has all been for our benefactor. With the new fiscal year, we start all over again—an even more ambitious goal supported by a budget that might get us halfway there, if we're frugal.

A definitive dataset, made with integrity and interpreted honestly, is the most innocent thing in the world, but whenever we produce one, Bill and I feel like Bonnie and Clyde celebrating yet another clean getaway. "In your face, universe!"

I shook my fists toward the ceiling on that night; then I ran my fingers through my stringy hair, trying to massage some fresh oxygen into my brain—a habit that I had picked up in graduate school. "You know, we're both getting too old for these long nights." I glanced at the clock and noted that my son had gone to bed several hours ago.

"But how shall we designate the apparatus?" Bill, energized by success, wanted to brainstorm over a funny name that could be condensed into an even funnier acronym. "I'm thinking we can work 'CAT' into it based on the nickel-catalyzed disproportion reaction."

No writer in the world agonizes over words the way a scientist does. Terminology is everything: we identify something by its established name, describe it using the universally agreed-upon terms, study it in a completely individual way, and then write about it using a code that takes years to master. When documenting our work, we "hypothesize" but never "guess"; we "conclude," not just "decide." We view the word "significant" to be so vague that it is useless but know that the addition of "highly" can signify half a million dollars of funding.

The scientific rights to naming a new species, a new mineral, a new atomic particle, a new compound, or a new galaxy are considered the highest honor and grandest task to which any scientist may aspire. Strict rules and traditions govern the naming conventions within each scientific field. You must muster all you know about what you've discovered and the world you live in, take what you remember and then figure out what makes you smile, make an allusion to something both contemporary and eternal, and finally christen the precious article as best you can, hoping against hope that some part of your clumsy label might stick through the ages to come. But on that night I was too brain-dead for the semantics-fest; I just wanted to go home and go to bed.

"We could call it 'four hundred and eighty thousand dollars of taxpayer money,' because that's what we spent making the damn thing," I suggested with a hiss toward the disobedient budget sheets that I was torturing toward reconciliation. I couldn't figure out who the hell to petition for more funding now that the project was over; we had maxed out all of our usual sources the previous year, and the budgets of every governmental agency that funded our research were shrinking. As much as I have loved being a scientist, I am ready to admit that I am tired of all the hard things that should be easy by now.

Bill watched me for a moment and then got up, slapping both thighs. "We don't have to call it anything. I'll just grind your last name into it. That's all it needs." We made eye contact and recognized fifteen years of our shared history reflected back in each other's eyes. I nodded my acknowledgment, and as I was still struggling to find the right words to thank him, Bill turned and walked out of my office.

He is strong where I am weak, and so together we make one complete person, each of us gaining half of what we need from the world and the other half from each other. I inwardly vowed to do whatever it took to raise more salary for him and to keep us going. As with many years before, I'd just have to find a way. Within two separate but adjacent rooms, we tuned two radios to different stations and went back to our work, having once again reassured each other that we are not alone.

2

LIKE MOST PEOPLE, I have a particular tree that I remember from my childhood. It was a blue-tinged spruce (*Picea pungens*) that stood defiantly green through the long months of bitter winter. I remember its needles as sharp and angry against the white snow and gray sky; it seemed a perfect role model for the stoicism being cultivated in me. In the summer I hugged it and climbed it and talked to it, and fantasized that it knew me and that I was invisible when I was underneath it, watching ants carry dead needles back and forth, damned to some lower circle of insect Hell. As I got older I realized that the tree didn't actually care about me, and I was taught that it could make its own food from water and air. I knew that my climbing constituted (at most) a vibration beneath notice, and that pulling branches off for my forts was akin to pulling single hairs off of my own head. And yet, each night for several more years, I slept ten feet away from that tree, separated only by a glass window. Then I went to college and began the long process of leaving my hometown, and my childhood, behind.

Since then, I have realized that my tree had been a child once too. The embryo that became my tree sat on the ground for years, caught between the danger of waiting too long and the danger of leaving the seed too early. Any mistake would surely have led to death, and to being swallowed up by a seething, unforgiving world capable of rotting even the strongest leaf in a matter of days. My tree had also been a teenager. It went through a ten-year period where it grew wildly, with little regard for the future. Between ages ten and twenty it doubled in size, and it was often ill prepared for the new challenges and responsibilities that came with such height. It strove to keep up with its peers and occasionally dared to outdo them by brazenly claiming the odd pocket of full sun. Focused solely on growth, it was incapable of making seeds yet prone to fits and starts of the necessary hormones. It marked the year as did the other teenagers: it shot up tall in the spring, it made new needles for the summer season, and it stretched its roots in the fall, until it reluctantly settled into a boring winter.

From the teenagers' perspective, the grown-up trees presented a future that was as stultifying as it was interminable. Nothing but fifty, eighty, maybe a hundred years of just trying not to fall down, unpunctuated by the piecemeal toil of replacing fallen needles every morning and shutting down enzymes every night. No more rush of nutrients to signal the conquering of new territory underground, just the droop of a reliable, worn taproot into last winter's new cracks. The adults grew a bit thicker around the middle each year, with little else to show for the passing decades. In their branches they stingily dangled hard-won nutrients above the perpetually hungry younger generations. Good neighborhoods, rich with water, thick soil, and—most important—full sunlight, give rise to trees that reach their maximum potential. In contrast, trees in bad neighborhoods never achieve half of that height, never have much of a teenage growth spurt, but focus instead on just holding on, growing at less than half the rate of the more fortunate.

During its eighty-odd years my tree was likely sick several times. Unable to run away from the constant barrage of animals and insects eager to dismantle it for shelter and food, it preempted attacks by armoring itself with sharp points and toxic, inedible sap. Its roots were the most at risk, smothered and vulnerable within a blanket of rotting plant tissue. The cost of maintaining these defenses came out of my tree's meager savings that were intended for happier uses: each drop of sap was a seed that didn't happen; each thorn was a leaf that wouldn't be made.

In 2013 my tree made a terrible mistake. Assuming that winter was over, it stretched its branches and grew a new crop of lush needles in anticipation of the summer. But then an unusual May brought a rare spring blizzard, and a copious amount of snow came down in just one weekend. Conifer trees can stand heavy snow, but the added weight of the foliage proved too much. The branches first bowed and then broke off, leaving a tall, bare trunk. My parents euthanized my tree by cutting it down and grinding out its roots. When they mentioned it on the phone months later, I was standing in the dazzling sunshine, living more than four thousand miles away in a place where it never snows. I think of the irony that I fully appreciated that my tree was alive only just in time to hear that it had died. But it's more than that—my spruce tree was not only alive; it had a *life,* similar to but different from my own. It passed its own milestones. My tree had its time, and time changed it.

Time has also changed me, my perception of my tree, and my perception of my tree's perception of itself. Science has taught me that everything is more complicated than we first assume, and that being able to derive happiness from discovery is a recipe for a beautiful life. It has also convinced me that carefully writing everything down is the only real defense we have against forgetting something important that once was and is no more, including the spruce tree that should have outlived me but did not.

3

A SEED KNOWS how to wait. Most seeds wait for at least a year before starting to grow; a cherry seed can wait for a hundred years with no problem. What exactly each seed is waiting for is known only to that seed. Some unique trigger-combination of temperature-moisture-light and many other things is required to convince a seed to jump off the deep end and take its chance—to take its one and only chance to grow.

A seed is alive while it waits. Every acorn on the ground is just as alive as the three-hundred-year-old oak tree that towers over it. Neither the seed nor the old oak is growing; they are both just waiting. Their waiting differs, however, in that the seed is waiting to flourish while the tree is only waiting to die. When you go into a forest you probably tend to look up at the plants that have grown so much taller than you ever could. You probably don't look down, where just beneath your single footprint sit hundreds of seeds, each one alive and waiting. They hope against hope for an opportunity that will probably never come. More than half of these seeds will die before they feel the trigger that they are waiting for, and during awful years every single one of them will die. All this death hardly matters, because the single birch tree towering over you produces at least a quarter of a million new seeds every single year. When you are in the forest, for every tree that you see, there are at least a hundred more trees waiting in the soil, alive and fervently wishing to be.

A coconut is a seed that's as big as your head. It can float from the coast of Africa across the entire Atlantic Ocean and then take root and grow on a Caribbean island. In contrast, orchid seeds are tiny:

one million of them put together add up to the weight of a single paper clip. Big or small, most of every seed is actually just food to sustain a waiting embryo. The embryo is a collection of only a few hundred cells, but it is a working blueprint for a real plant with root and shoot already formed.

When the embryo within a seed starts to grow, it basically just stretches out of its doubled-over waiting posture, elongating into official ownership of the form that it assumed years ago. The hard coat that surrounds a peach pit, a sesame or mustard seed, or a walnut's shell mostly exists to prevent this expansion. In the laboratory, we simply scratch the hard coat and add a little water and it's enough to make almost any seed grow. I must have cracked thousands of seeds over the years, and yet the next day's green never fails to amaze me. Something so hard can be so easy if you just have a little help. In the right place, under the right conditions, you can finally stretch out into what you're supposed to be.

After scientists broke open the coat of a lotus seed (*Nelumbo nucifera*) and coddled the embryo into growth, they kept the empty husk. When they radiocarbon-dated this discarded outer shell, they discovered that their seedling had been waiting for them within a peat bog in China for no less than two thousand years. This tiny seed had stubbornly kept up the hope of its own future while entire human civilizations rose and fell. And then one day this little plant's yearning finally burst forth within a laboratory. I wonder where it is right now.

Each beginning is the end of a waiting. We are each given exactly one chance to be. Each of us is both impossible and inevitable. Every replete tree was first a seed that waited.

4

THE FIRST TIME I performed an experiment that wasn't a rote classroom exercise I was nineteen, and I did it because I wanted the money.

I must have held ten different jobs while I was an undergraduate at the University of Minnesota in Minneapolis. For the entirety of the four years that I was there, I worked twenty hours a week and much more than that during the breaks, earning money to supplement my scholarship. I worked as a proofreader for the university's press, a secretary to the dean of agriculture, a cameraperson for the long-distance learning program, and a machinist polishing glass slides. I taught swimming lessons, fetched library books, and ushered rich people to their seats within Northrop Auditorium. But none of it compared with the time that I spent working in a hospital pharmacy.

A girl from my chemistry class recommended me for a job in the university hospital, where she worked. The money was good, she said, and they let you work two eight-hour shifts in a row, earning time and a half during the second shift. Her boss hired me upon introduction, and after a disconcerting lack of scrutiny toward my credentials, I found myself the proud possessor of two new sets of turquoise scrubs.

The next day I showed up after classes at two-thirty in the afternoon, ready to work the three-to-eleven shift. I would be working in the basement of the hospital, within the main pharmacy that housed, sorted, and generally kept track of every medication administered to every patient in the hospital. It was a huge facility all its own, with an information desk, a delivery dock, and several storage rooms, including cold lockers kept at a range of low temperatures. It was built around an open-floor laboratory that seemed the size of a warehouse, full of people mixing the made-to-order prescriptions needed for the more complex therapies that were happening all over the hospital. The deputy pharmacist-in-charge explained to me that I would start as a "runner," hand-delivering intravenous pain medications to the nursing stations where they were needed.

In those days, a doctor had to request medication by writing a paper prescription, which was then personally escorted to the hospital pharmacy. Within the pharmacy laboratory, a tiny amount of pure painkiller was injected into a floppy bag of fluid and immediately swaddled in a thick cloak of paperwork that required both a signature and a time stamp every time the bag was transferred between hospital workers. After the formula and the amount of drug it contained were double-checked and signed off on by a professional pharmacist sporting a Pharm.D. degree (the pharmacist's equivalent to the doctor's M.D.), the bag was handed to a runner, who signed it and then walked the bag through the hospital and handed it directly to the nurse in charge of the patient, who signed it again and then presumably administered the treatment.

After drop-off, it was the runner's duty to check the station's outbox for any other M.D. orders and, upon finding any, to carry them back to the pharmacy. It thrilled me that my signature was required to advance what I saw as an urgent process, and this ushered in a rich fantasy life, within which I was routinely relieving suffering, saving souls, and generally preserving the dignity of life all around. Just like every girl who has ever once earned an A in a science course, I had been encouraged to go to medical school, and I began to consider it, hoping against hope for a whopping scholarship of some kind.

My job as a runner allowed me to roam every hallway of the hospital and learn the idiosyncrasies of how each nursing station worked, although the majority of my trips simply navigated the well-traveled route between the pharmacy and the hospice ward. I learned to work through long hours that were unbroken by social interaction, save a signature and a look. Although I was surrounded by other people, never-dimming light, and humming machines, I worked in isolation and was no more disturbed by all the activity than I was by my own breathing.

I also discovered that I could set my subconscious to work on a specific task while my conscious mind was occupied with performing workday rituals. During my interview for the job, I had looked longingly into the main lab, where so many technicians worked industriously, filling and injecting needles, scrutinizing vials, and unwrapping sterile tubes. I had asked the pharmacist what they were making. "Mostly antiarrhythmics, heart-attack stuff," she explained.

The next morning I told my English professor that the topic of my term paper would be "The Use and Meaning of 'Heart' Within *David Copperfield.*" The high that I was riding after my declaration began to wear off as I immersed myself in flipping the pages back and forth, cataloging what turned out to be hundreds of instances of "heart," "hearty," "heartfelt," and other derivatives within the first ten chapters of the book alone. I decided to restrict my paper to a treatment of only the most significant instances. This approach backfired when I hit the following within chapter thirty-eight: *What I cannot describe is, how, in the innermost recesses of my own heart, I had a lurking jealousy even of Death.* I thought about it and thought about it, but I didn't get anywhere. Finally, two o'clock rolled around and I had to go into work.

That night in the hospital I walked in and out of the hospice ward ten or twenty times, and my eyes and hands moved through the necessary tasks. Well into the night and deeper in my brain, it came to me that as hospital workers, we were being paid to trail along behind Death as he escorted frail, wasted bodies over difficult miles, dragging their loved ones along with him. My job was to meet the traveling party at its designated way stations and faithfully provide fresh supplies for the journey. When the weary group disappeared over the horizon, we turned back, knowing that another agonized family would be arriving soon.

The doctors, nurses, and I didn't cry because the bewildered husbands and stricken daughters were crying enough for all of us. Helpless and impotent against the awesome power of Death, we nonetheless bowed our heads in the pharmacy, injected twenty milliliters of salvation into a bag of tears, blessed it again and again, and then carried it like a baby to the hospice and offered it up. The drug would flow into a passive vein, the family would draw close, and a cup of fluid might be temporarily removed from their ocean of pain. When my hospital shift ended I found that I could go home and write pages and pages, while in contrast, sitting in front of the computer for hours before work had yielded nothing. I memorized the difficult passages from the book, and then let my subconscious work out their meaning while I was at the hospital.

Every hospital worker was required to take three twenty-minute breaks during each eight-hour shift, but runners were also requested to coordinate and stagger them, so that if things got unexpectedly busy, we would be down only one person at any given time. This policy forced me to manage the timing and extent to which I could wander through my own thoughts, and I developed a fine control over my ability to reemerge. I could work with my brain in my hands for hours, move it into my head for twenty minutes, and then shunt it back into my fingers in the same way that I could slosh water back and forth in a half-full bucket.

I spent my breaks in the tiny courtyards that nestled among the hospital's buildings, luxuriating in the natural light and unrecycled air. One morning I was lying on the grass with my legs elevated, counting the cigarette butts on the ground while trying to drain some blood back into my upper body. *The early sun was striking edgewise on its gables and lattice-windows, touching them with gold; and some beams of its old peace seemed to touch my heart,* I recited from chapter fifty-two. I saw my supervisor peer around the courtyard wall and then motion for me to come inside. For a moment I was terrified that I had lost track of time, until my watch told me that my break still had five minutes left in it. When we got back to the pharmacy lab, both my supervisor and the Pharm.D. were looking at me seriously. "Why don't you leave out the front door when you run a controlled-sub bag?" one of them asked me.

"Because I go and use the back stairway."

"But you can't get to the walkway that crosses over to the hospice tower from those stairs," my supervisor argued.

"You can if you cut through the cafeteria loading dock."

"So you never use the elevator?" The Pharm.D. looked puzzled.

"It's a shortcut, and you never have to wait," I answered. "It really is quicker, I've timed it. And, I mean, there's somebody up there in pain, waiting for this stuff, right?" My supervisors looked at each other, rolled their eyes, and returned to their other tasks.

It was only partly true that I had developed this route in order to cut my delivery times. I was also compelled to keep moving in order to burn off the endless energy that I possessed at that age, which seized me in ferocious spurts, keeping me awake for days at a time. Hospital work gave me a place to be and a mission to accomplish, complete with repetitive tasks that helped to constrain my racing thoughts.

Near the end of the afternoon shift, it never failed that some runner would call in sick, giving me the option to work the night shift if it looked as if I wasn't going to sleep anyway. Sooner or later I would go home at least exhausted if not sleepy, comforted by the night's companionable wakefulness, and a full paycheck richer in the bargain; *the shells and pebbles on the beach . . . made a calm in my heart,* I remembered from chapter ten. Plus I was doing something important—or so I had convinced myself.

About a month after I had explained my delivery routes, I walked into the pharmacy and the Pharm.D. turned and called out, "Lydia, she's here!" She then turned to me and explained, "Lydia is going to train you to shoot bags," and just like that, my career as a runner was over. Lydia stood up from her post and cast a sideways glance at the boss. The look on her face didn't exactly congratulate me on my promotion, but I sensed that Lydia's training was a necessary hurdle positioned between me and a big fat raise.

"Come on back and take a load off!" Lydia yelled in her gravelly voice, more to annoy the Pharm.D. than to welcome me. I got excited and *my heart leaped with a new hope of pleasure,* to quote chapter twenty. Instead of delivering the bags of intravenous medications, I was going to make them, and then hand them over to someone else for double-checking and delivery. I pictured myself sitting down at my own workstation and adjusting my stool to the perfect height. I pictured myself walking importantly back and forth from the stock table, picking out exactly the right little bottles of concentrated drugs in the same way that a rich woman confidently seizes upon the perfect shade of nail polish before her manicure. I saw myself take my position, sit up straight, square my shoulders, and begin to work my magic, calmly but quickly—because, after all, someone's life was at stake.

"Here, pull your hair back tight"—Lydia interrupted my daydream by dangling an industrial rubber band in front of my face—"and you better get used to yourself without makeup. I bet you think that I look like this hell all the time," she mused with a grim smirk. Loose hair, nail polish, jewelry—none of these were allowed in the pharmacy, as they presented additional surface area that might harbor contamination, and so I adopted the frazzled "natural" look that you so often see in hospital staff, and I have kept it to this day.

The employee pool was about evenly split between preprofessional students and career technicians, but I didn't fit in with any of them. Like the students, I had classes and exams to worry about, but like the techs, I worked way too much, because I just needed somewhere to be. Lydia was what was called a "lifer" within the pharmacy laboratory, which meant she was already there when anyone queried had been hired. While I stowed my backpack, Lydia notified our supervising Pharm.D. that she was going to teach me the differences between the drugs in the stockroom, where they were shelved according to their chemical formulas. I was only mildly surprised when she walked right past the stockroom and toward the courtyard instead.

Lydia was famous for two things: her breaks and her rides. She would take all of her breaks during the first ninety minutes of each of her eight-hour shifts and attempt to smoke the three packs of cigarettes that she would have consumed across eight hours had they been leisure time. Smoking sixty cigarettes in sixty minutes requires no small amount of concentration, and even though it was easy to locate her in the courtyard, she was generally too occupied for conversation. During the second hour of her shift, Lydia was extremely alert and productive, but it was generally wisest to give her a wide berth about five hours later, when any perceived slight could really set her off. During the last twenty minutes of her shift, even the pharmacists avoided her as she sat stiffly watching the clock, clutching a sterile needle in her shaking fist.

It seemed out of character, but if you were female and worked a shift with Lydia that ended at night, she would insist on giving you a ride home. An incoherent string of snarls about "asshole rapists" was the only explanation that we ever got for her strategic generosity. It was hard for me to picture these rapists, dressed for the twenty-below weather and circling the hospital in a holding pattern until 11:00 p.m., when a fatigued herd of nursing-student prey would stagger into their hunting grounds, but in that part of the country we didn't have the kind of Januaries where you turned down a ride for any reason.

Once released from the gas chamber of secondhand smoke that also served as Lydia's car, you had to strip off your scrubs in the hallway in order to prevent your apartment from smelling like the coal miners' union hall for a week. Lydia never drove off until she saw you go inside and flash the porch light on and off. "Blink it more than once if somebody needs their balls ripped off," she instructed us maternally. *She did not replace my mother; no one could do that; but she came into a vacancy in my heart, which closed upon her,* I remembered from chapter four, and smiled to myself.

When—during my first hour in the pharmacy laboratory—Lydia and I got to the courtyard, we sat down in the metal chairs at one of the outdoor tables. She pulled a pack of Winston Lights out of the fold of her sock and thumped it three times on the heel of her hand. She slid the pack over to me as an offering, and lit her cigarette with the communal lighter that was kept chained to a branch of a little birch tree that was serving hard time planted in cement. Lydia put her feet up and took a long drag with her eyes closed. I played with her pack of cigarettes, first shaking them out and then reloading them, although I didn't smoke.

In my eyes, Lydia was ancient, which meant that she was probably about thirty-five years old. At least thirty-four of those years had been hard years, I figured, given the way that she carried herself. I supposed that Lydia had also been unlucky in love, since she perfectly fit the type who would be sitting at a kitchen table nursing a coffee cup full of gin while waiting for the kids to come home from school, had she been blessed within a matrimonial union. Chapter thirty-six said it better than I could: *She gave me the idea of some fierce thing, that was dragging the length of its chain to and fro upon a beaten track, and wearing its heart out.*

To my surprise, Lydia was also curious about me and asked me where I was from. After learning the name of my hometown she responded, "Yeah, I've heard of it; that's where that big hogkill is at. Cripes, you crawled right out of the armpit of the Midwest, didn't you?" I shrugged, and she continued, "Well, there's only one place worse than that, and it's the frozen shithole up north where I grew up." She threw a smoldering butt on the ground, looked at her watch, and then lit another cigarette.

We passed the next five minutes in silence. Finally, she exhaled and said, "You ready to go back?" I shrugged by way of answer, and we both stood up. "You just do what I do, okay? I'll go slow, it'll be fine," she said, and thus concluded my formal training in pharmaceutical medicine. I still didn't have a crystal-clear idea of how I was to assemble a sterile mixture of medications that could be injected into a desperately sick person's vein, but I guessed that I would pick it up as I went along.

Sitting next to Lydia and carefully mimicking her actions didn't turn out to be a bad way to learn sterile technique, which is more like dancing with your hands than it is like making something. The air through which we walk, both outdoors and inside our buildings, contains plenty of tiny organisms that would feed quite happily on our insides but don't usually bother us because they can't get close enough to our juicy parts, such as our brains and hearts. Our outer skin is thick and whole, and any openings, such as those for our eyes, nose, mouth, and ears, are coated in protective slime and wax.

This also means that every needle in every hospital might be the winning lottery ticket for a lucky random bacterium who, after recovering from the initial rush of injection, finds himself swishing along within a jolly river of blood until he disembarks in some quiet cul-de-sac of the kidneys, perhaps. There he will breed and also produce one bumper crop after another of toxins that are all the harder for us to fight because they were produced near our organs. The bacteria represent only one hostile faction, with viruses and yeasts capable of their own similar modes of destruction. A sterile needle represents our best defense against such an onslaught.

When a nurse gives you a shot, or draws your blood, it's a relatively quick puncture, in and out, and afterward your skin closes over and reestablishes multiple firewalls against reentry. Your caregiver ensures against bacterial stowaways by using a syringe with a pointed tip that's been sterilized and then sealed into a protective plastic cap. She rubs your skin with rubbing alcohol (isopropanol) in order to cleanse your outermost layer of any bacteria that might otherwise get shoved into your body during the injection.

When you are given an intravenous medication, it's a little different. The nurse cleans your skin, inserts a needle, and then leaves it there for hours, effectively making the needle, the tube, and the entire bag that is attached to it an extension of your vein—and all the liquid in the bag becomes an extension of your bloodstream. She will hang the bag over your head in order to encourage fluid to flow from the bag into you and not vice versa, and she'll activate a pump to very gently force it in if the doctor recommends it. The entire contents of the bag will mix with your bloodstream, and any excess from the two pools will be stored in the overflow chamber that is your bladder.

Under this configuration a bacterium now has a lot more territory in which to ready itself for action. It's not just the tip of the needle that could harbor infection; now it is the entire inner surface of the bag and tubes—not to mention the fluid itself—an expanse more than one hundred times larger than that of the syringe alone. Of course this means that the entire apparatus must be kept sterile, but it also means that everything that has ever touched the whole mess, while the medications were added and mixed—and even before that, when the chemical ingredients were synthesized and stored—must be kept sterile at each step along the way.

The great thing about an IV is that it allows your doctor to deliver medicine to your body quickly, and for a sustained period. During cardiac arrest, your brain doesn't have a couple of hours to wait around for oxygen while your heart stands in a long queue behind your stomach and intestines, hoping for its share of the medication to seep out of the pill you somehow swallowed. So how to combine a liter of fluid with active agents, customized according to the patient's weight and status, while keeping everything sterile? If this is for the ER or the ICU, we have about ten minutes to make it happen. Fortunately for the patient, there is a sleep-starved teenager apprenticed to a chain-smoking barmaid in the basement who is ready for action.

* * *

The first step is to create a clean workspace. Although it's hard to picture, bacteria, yeast, and other tiny things can be removed from air by forcing it through a mesh with holes that are three hundred times smaller than the diameter of a human hair. When I make intravenous medications, I sit in front of a wall that blows air through the mesh and toward me. The area between me and the wall, therefore, is a clean space where sterile items can be opened, mixed, and resealed.

The first thing I do after I pull on gloves is to liberally wash down my entire work area using a spray bottle of isopropanol, bathing the counter and my gloved hands over and over and wiping with tissue after tissue. I leave all surfaces damp with isopropanol, knowing that the sterile air stream will dry them, mainly by blowing it straight into my face.

I go to the Teletype and select a medication order in the form of a two-inch-by-two-inch sticker, upon which is typed the patient's name, gender, and location and a code that specifies the mixture of medications required. I pick out a sealed bag of fluid, which has the approximate shape and feel of a packaged pork loin, from the pile produced by the tech who is "pumping bags," filling them by the liter with either normal saline or Ringer's solution—a weakly sugared saline named after Sydney Ringer, who in 1882 found that he could make a dead frog's heart beat by repeatedly bathing it in this very formula. Reading the medication order, I pick up the bag, peel the backing off of the sticker, and attach it to the top side such that the text will be upside down when the bag hangs and drips into the patient.

I carry the bag to the stock table and pick up the concentrated drugs that I will need, and I restock my own personal stash of the ones that I use very frequently. These drugs come in little bottles with rubber-stoppered tops, color-coded for quick recognition. The tops are crimped shut with aluminum, and the glass and metal sparkle in the unrelenting brightness of the lit laboratory. Some of the bejeweled bottles are indeed precious, containing only the tiniest drop of liquid protein concentrated from the bodies of heroic human donors or hapless animal subjects. Each of these miniature glittering bottles contains a day, or perhaps a week, of frustration for a ruthless tumor—perhaps just long enough for an ill-remembered animosity to thaw into a critical goodbye, or so I fantasize while I work.

Returning to my workstation, I place my materials in front of me, in a straight line across the front of the bench. I set the bag of fluid that I will inject down on my left, careful to point the site of injection toward the panel of blowing air. The upside-down sticker is now facing toward me so that I can read it, and I widely space the medications from left to right, in the order that I will inject them into the bag. Beside each bottle I place a syringe sized to accommodate the amount of the drug specified on the sticker. I double-check the entire setup, left to right, comparing the words on the sticker with the words on the bottles, one after the other, the first three letters of each word only in order to prevent wasting the time it would take to read the whole name.

I take a deep breath and grab a stack of alcohol wipes, the kind ever so slightly foiled within their tear-open package, as is my preference. I steady my hands, reach around the bag, and pull the seal off of the injection port that faces away from me. I raise an alcohol wipe in front of me, tear it open, and bring it down in front of the bag. I clean the rubbery port that the needle is to penetrate and swab the wipe up and back, careful not to let my hands pass between it and the blowing wall. Then I clean the first bottle of medication the same way, using a different wipe.

While turning the small vial of medication upside down with my left hand, I pop the cover off of the syringe with my right hand. I clutch the items securely but strangely in order to keep my fingers to the back, as if I was exposing each item to some holy light. I draw the exact amount of medication printed on the sticker into the syringe, making sure that my eyes are level with the fluid line so that I do not misread the number of milliliters that was measured. I pull the bottle up and off of the syringe by flexing the muscles in my left hand, careful to simultaneously relax the muscles in my right hand in order to avoid losing a drop of medication out of the tip of the needle during separation.

I set the bottle down carefully and move the needle up and over the front of the bag; then I inject the medicine into the bag and toward me. I move the needle up and out, and it is instantaneously useless. I position the syringe's plunger back at the level of medication that I injected and set it down empty on a tray outside my workstation. I carefully seal the bottle of medication that I just injected and then place it on the tray just to the right of its syringe. I do this until I have used each bottle, and thus completed the recipe. Then I carefully reseal the bag with a plastic cap and lay it across the same tray, on the side facing away from the needles.

I take off my gloves, pick up a pen, and sign my initials in one corner of the sticker on the bag, assuming partial responsibility for I-don't-know-what. I place the tray in the queue that is serviced by a senior Pharm.D., who methodically double-checks every label, every syringe, and every bottle to ensure that the bag contains what was ordered. If a mistake is found, the bag is discarded, the sticker is reprinted, the whole thing is now a rush job, and a lifer intercedes.

It doesn't matter that this is my first day in the laboratory. There are no practice bags. There is just doing it right or doing it wrong. While we work we are watched to make sure that we don't preferentially draw out the simpler orders from the Teletype, and that we use up the entirety of each bottle of medication before we open a new one. We are constantly reminded that any mistake we make could kill someone. The number of medication orders far exceeds what we can complete by the time they are needed, and we are constantly behind. The more people who call in sick, the fewer of us are in the lab working, the faster we have to work, and the further behind we get.

There is no time to discuss the fact that this horrible, horrible system is not working, or to assert that we are neither criminals nor machines. There are only endless medication orders, given by other exhausted people with nobody better than us to depend upon.

Working in the hospital teaches you that there are only two kinds of people in the world: the sick and the not sick. If you are not sick, shut up and help. Twenty-five years later, I still cannot reject this as an inaccurate worldview.

* * *

Lydia was magnificent at her workstation, possibly because she'd been doing this sixty hours a week for almost twenty years. Watching her sort, clean, and inject was like watching a ballerina defy gravity. I watched her hands fly and thought . . . *in an easy amateur way, and without any book (he seemed to me to know everything by heart),* from chapter seven. On that first day I witnessed her shooting at least twenty bags, sometimes with her eyes closed. I never saw her make a mistake. I was certain that she worked while in some kind of trance, as there was no way that her brain could have been sufficiently oxygenated. One of the worst things one can do is sneeze or otherwise spray bodily fluids into a sterile space, and Lydia, for whom the very act of exhaling was basically a cough, exhibited breath control that was positively superhuman while mixing medications.

Within my first couple of hours at my workstation, I had successfully made a few bags of simple electrolytes, and the supervisor had started pressuring me to pick up and shoot some of the more difficult orders because the lab was running severely behind. I tried an order for a simple "benzo bag" but then panicked after I injected the sedative, knowing that if I had somehow injected more than I thought I did, I could be curing the patient's anxiety with much more finality than anyone expected. Terrified as a trapped animal, I actually considered bluffing my way through it, putting the bag on the tray in the queue and then moving on with my life. But I came to and all at once realized just how crazy that instinct was. I took the bag over to the sink, sliced it with a scalpel, and dumped its contents down the drain while the Pharm.D. gave me the evil eye. I walked back to Lydia and suggested that we take a break.

"I don't think I can do this," I confessed when we got to the courtyard. "This is the most stressful thing I've ever done."

Lydia chuckled. "You're making way too much out of this. Remember, it ain't brain surgery."

"Yeah, brain surgery's on the fifth floor." I completed the joke that the runners told each other at least five times a day. "Still, what if I just can't get it?" I groaned. "Half the time I don't remember if I did something right or wrong."

Lydia looked around and then leaned forward and spoke. "Listen, I'm gonna tell you something about sterile technique." She leaned back and continued in a low voice. "Now, don't go licking the needle or anything, but if you've got something on your hands that's gonna kill them—they're gonna die anyway." I had no answer for that, and Lydia seemed to think that she'd explained what needed to be explained, and so we sat in silence while she smoked.

After a while I rubbed my temples and said, "Man, I've got a headache. Lydia, don't you ever worry about what breathing all this alcohol is doing to our lungs?"

Lydia had a cigarette hanging out of her mouth at the time, and the look she gave me showed that she now had proof that I was terminally stupid. She took a long, long drag and then answered while exhaling. "What do you think?"

Right after we got back from break, I threw myself into the fray and drew out an order for a complicated chemotherapy bag, determined to make good on what was left of my first day in the laboratory. I made the bag accurately and was very proud of myself until an enraged Pharm.D. walked up to me and held a tiny bottle of precious interferon an inch from my face.

"You just *wasted* this entire bottle," she hissed with fury. A few minutes before that, I had injected the valuable immunity promoter and then removed the bottle from my workspace without sealing it, thus contaminating the remainder of the supply. In one stroke I had wasted at least a thousand dollars of the medication, and also created a big paperwork problem. I felt a rush of shame such as I hadn't experienced since I was a little girl getting caught reading on the wrong page by yet another teacher who was sick and tired of dealing with me. As the protagonist of my own private chapter seven, *I looked up with a flush upon my face and remorse in my heart*.

Lydia, smelling opportunity, stepped in and assured the steaming Pharm.D., "She just needs a break—she hasn't had a break all day. C'mon, kiddo, let's go." She led me to the courtyard, and we commenced our umpteenth break for that very shift.

When we got there, I sat down and put my head in my hands. "If I get fired, I don't know what I'll do," I said, hiccupping as tears threatened.

"Fired? Is *that* what you think?" Lydia cackled. "Jesus, relax. I've never seen anybody get fired from this hellhole. In case you haven't noticed, long before people get fired, they quit."

"I can't quit," I confessed with anguish. "I need the money."

Lydia lit a cigarette and took a long drag while looking at me. "Yep," she said sadly, "you and me are the type that can't quit." She waved her pack of Winston Lights toward me and I declined for the sixth time that day.

When Lydia dropped me off at my apartment later that night, I asked her what she sat and thought about during those long hours when we worked silently in the pharmacy.

She considered the question for a moment, and then answered, "My ex-husband."

"Let me guess," I ventured, "he's in prison?"

"He *wishes,*" she snorted. "Bastard lives in Iowa."

As we sat there and laughed at a joke that's as old as Minnesota itself, chapter seven echoed in my head: *miserable little dogs, we laugh, with our visages as white as ashes, and our hearts sinking into our boots*.

When the medication orders slowed down in the pharmacy and I was tired of sitting, I would go and visit the blood bank to see if they wanted me to carry some pints over to the emergency room. This afforded me the opportunity to burn off some energy, since there was always a lot of time for pacing while waiting for the number and type of blood units to be verified repeatedly by all parties.

The lifer who worked the three-to-eleven shift behind the counter was named Claude, and while not as ancient as Lydia, he still qualified as a senior citizen in my eyes, having arrived at the ripe old age of twenty-eight. Claude fascinated me, both because he was the only person I'd ever met who had been to jail and because he was easily one of the most harmlessly nice guys I'd ever known. His hard life had left him physically the worse for wear, but he didn't seem to harbor any resentment, and I supposed this was because his attention span was so short and shallow. Working the blood desk was hands down the easiest job at the hospital, Claude had boasted to me with a sort of confused pride.

Claude explained that there were only three things that he had to remember how to do: thaw blood, check blood, and dump blood. Claude began each shift by wheeling several pallets, each stacked high with bricklike bags of frozen blood, out of the deep freeze and into the plus-five-degree room to thaw. Immediately after it was donated and processed, the blood had been frozen and stored, and now it had to thaw slowly in order to be usable. By moving the blood, Claude was preparing the stash that would be available for use three shifts hence. The next thing Claude did was to man the counter for about seven hours, waiting for someone to come down with a blood order. Before signing anything out, he had to check and double-check the blood type on the bags against the order form, and sometimes call the operating theaters to double-check. He explained to me that there were "at least four or six" different types of blood and that sending off the wrong type "could waste it by killing somebody," and it troubled me that the two consequences were not entirely discrete within his mind.

When Claude slammed down the limp yellowy bags of blood plasma in bundles of three, I couldn't help but think of the butcher shops that lined the Main Street of my hometown, and particularly of the meat counters where Mr. Knauer whumped down whatever my mother's note requested before sending me back home to help administer dinner to my family. Near the end of his shift, it was Claude's job to discard any unused thawed bags, which amounted to gallons and gallons of blood, down the hazardous-materials chute, where they would be incinerated along with the rest of the day's medical rubbish. It seemed a waste to me, and I commented on what a shame it was that good-hearted citizens had gone out of their way to donate the blood that he was heaving by the armload into the Dumpster.

"Don't feel too bad," Claude said with genuine sensitivity. "It's mostly just bums doing it for the cookie."

The guys at the blood bank were infamous for hitting on pharmacy runners, so I wasn't particularly flattered when Claude developed a crush on me. "When I heard that bunch of ambulances come in I started hoping that I would see you down here," he told me one day when I arrived with an order, prompting me to mention my fictitious art-student boyfriend, whom I had mentally rendered in detail for use on just such occasions.

"If you've got a boyfriend, why are you working here?" Claude asked, and it dawned on me that his understanding of the relationship between the sexes was undoubtedly deeper than mine. I made the excuse that artists are generally penniless, even when they are gorgeous and wear a sort of troubled look strikingly reminiscent of a certain photograph of Ted Williams at bat during the 1941 All-Star Game.

"Oh, so he needs you to buy pot for him," Claude said with what might or might not have been sarcasm, and I couldn't think of a comeback with which to defend my imaginary boyfriend, so I let it slide.

I took to working the 11:00 p.m. to 7:00 a.m. shift and made a point of being there on Tuesday and Thursday mornings, to shoot and then deliver a cart of "drop bags" to the psychiatric ward. These were saline-based intravenous medications containing a sedative called droperidol, to be used as anesthesia during electroconvulsive therapy, known by caregivers as "ECT" and misunderstood by the public to be "shock treatments." Twice a week, patients were readied in the early-morning hours and settled onto gurneys, and then lined up in the hallway to wait their turn for the procedure. One by one, each patient was drawn into a quiet room where a team of doctors and nurses administered electrical stimulation to one side of the head while carefully monitoring vital signs; all the while the patient was awash in the anesthesia I had brought.

Accordingly, Wednesdays and Fridays were noticeably better days in the ward, when many of the patients who had previously seemed dead in all but body could be found sitting up, dressed in their street clothes. Some could even briefly look me in the eye. In contrast, Sundays and Mondays were the worst days in the ward, when patients rocked and scratched themselves or moaned while lying in bed, cared for by nurses who seemed both supremely capable and acutely helpless.

The first time that I entered through the double-locked doors of the psych ward I was terrified, believing for no real reason that such places harbored evil souls ready to assault me at any moment. But once inside I found it to be the slowest-moving place on Earth, and I saw that these patients were unique only in that time had stopped inside their wounds, which were seemingly never to heal. The pain was so thick and palpable in the psych ward that a visitor could breathe it like the heavy humidity of summer air, and I soon realized that the challenge would not be to defend myself from patients, but to defend myself against my own increasing indifference toward them. What originally struck me as cryptic in chapter fifty-nine was now mundane: *they are turned inward, to feed upon their own hearts, and their own hearts are very bad feeding.*

After a few months in the hospital lab, I became really good at shooting bags, to the extent that I could keep up with Lydia and even outstrip her at times. Eventually the Pharm.D.'s double-checking quit turning up errors in my work, and soon afterward, my confidence ripened into boredom. I challenged myself by developing time-saving rituals for everything from how I lined up my medications to the number of steps I took while walking to the Teletype. I studied the names on each label and began to recognize the sicker patients who required the same mixtures day after day. I started shooting the tiny bags that required complicated dilutions, made for infants born prematurely and bearing stickers that read only "Baby Boy Jones" or "Baby Girl Smith" where there would otherwise have been a full name.

Occasionally I was handed a "cut slip," printed from a second, quieter Teletype, which informed the pharmacy when a patient requiring medication had died, so that the order was no longer needed. If a Pharm.D. tapped me on the shoulder and presented me with a cut slip, I stood up, walked to the sink, slashed or "cut" the bag that I was shooting and poured it out, and then grabbed a new order on the way back to my chair. One day I got a cut slip for a chemotherapy bag that I was making for a patient whose name I was in the habit of searching out from the pile daily. I stopped and looked around. Somehow I felt I had a simple sort of respects to pay, but who would want them?

Slowly I went from believing that I was doing the most important work in the world to ruminating over how pointless it was to be part of a pharmaceutical chain gang producing a mule train of medications to be hauled upstairs every hour of every day forever without end. From this darker perspective, the hospital was a place where you confined a sick person and then pumped medication through him until he died or got better, and it was not more complicated than that. I couldn't cure anybody. I could follow a recipe and wait to see what happened.

Just as I reached the peak of my disenchantment, one of my professors offered me a long-term work-study position in his research laboratory, and all at once I was assured of the money that I needed to keep me in college until I got my degree. So I quit my hospital job and gave up on saving other people's lives. Instead, I started working in a research laboratory in order to save my own life. To save myself from the fear of having to drop out and from then being bodily foreclosed upon by some boy back home. From the small-town wedding and the children who would follow, who would have grown to hate me as I vented my frustrated ambitions on them. Instead, I would take a long, lonely journey toward adulthood with the dogged faith of the pioneer who has realized that there is no promised land but still holds out hope that the destination will be someplace better than here.

On the same day that I gave my notice to the human resources office at the hospital, I sat through my break with Lydia. While she smoked, she explained to me that I should never buy a Chevrolet because they just wouldn't run reliably for a woman driver. She had always stuck with Fords and had yet to have one completely crap out on her. During a pause, I shared with Lydia that I'd gotten a better-paying job and that I was quitting the pharmacy. True, I'd been working in the hospital for barely six months, but I had begun to see it clearly for what it was. This place was a hellhole, I had realized, just as she had been telling me since the first day I met her.

I augured grandly that someday I would have my own research lab that would be even bigger than the one I was leaving for, and I wouldn't hire anyone who didn't care as deeply about the work as I did myself. I completed my speech in a little crescendo of self-importance from chapter ten: it was inevitable that I would be *working with a better heart in my own house . . . than I could in anybody else's now.*

I knew that she had heard me, and so I was surprised when she just looked away and took a drag on her cigarette instead of responding. After a moment, she tapped the ash off and continued talking about cars, picking up exactly where she had left off. After we both got off work at 11:00 p.m., I waited around for a bit, but then started toward my apartment on foot.

It was a clear night and so cold that the snow squeaked underneath my feet as I walked. After I had gone a few blocks, Lydia's car passed me while I was trudging along and I was stung with a new type of loneliness. *The old unhappy loss or want of something had, I am conscious, some place in my heart* came to me from chapter forty-four. I watched Lydia's single functioning taillight disappear across the bridge, lowered my head against the wind, and continued to make my own way home.

5

NO RISK IS MORE TERRIFYING than that taken by the first root. A lucky root will eventually find water, but its first job is to anchor—to anchor an embryo and forever end its mobile phase, however passive that mobility was. Once the first root is extended, the plant will never again enjoy any hope (however feeble) of relocating to a place less cold, less dry, less dangerous. Indeed, it will face frost, drought, and greedy jaws without any possibility of flight. The tiny rootlet has only one chance to guess what the future years, decades—even centuries—will bring to the patch of soil where it sits. It assesses the light and humidity of the moment, refers to its programming, and quite literally takes the plunge.

Everything is risked in that one moment when the first cells (the "hypocotyl") advance from the seed coat. The root grows down before the shoot grows up, and so there is no possibility for green tissue to make new food for several days or even weeks. Rooting exhausts the very last reserves of the seed. The gamble is everything, and losing means death. The odds are more than a million to one against success.

But when it wins, it wins big. If a root finds what it needs, it bulks into a taproot—an anchor that can swell and split bedrock, and move gallons of water daily for years, much more efficiently than any mechanical pump yet invented by man. The taproot sends out lateral roots that intertwine with those of the plant next to it, capable of signaling danger, similar to the way that information passes between neurons via their synapses. The surface area of this root system is easily one hundred times greater than that of all the leaves put together. Tear apart everything aboveground—everything—and most plants can still grow rebelliously back from just one intact root. More than once. More than twice.

The deepest-growing roots are those of the gutsy acacia tree (genus *Acacia*). When the Suez Canal was first dug, the thorny roots of a scrappy little acacia tree were found extending twelve meters, forty feet, or thirty meters downward, depending on whether you are reading Thomas (2000), Skene (2006), or Raven et al. (2005), respectively. I suspect that the authors of these botany textbooks included the Suez Canal anecdote in order to teach me something about hydraulics, but the story has left me with a dank and dusky false memory instead.

In my mind it is 1860, and I see a ragged cohort of men stumble upon a living root while they are digging more than one hundred feet belowground. I see them stand gaping in the fetid air, slowly overcoming their disbelief that this root could somehow be attached to some tree that is growing far above them. In fact, both parties registered their disbelief that day: the acacia tree was also undoubtedly surprised to find its roots exposed from the rock that confined it, and produced a flood of hormones in response, first locally and then eventually diffusing through every cell of its being.

When those men moved soil and rock in order to form an unprecedented path between the Mediterranean and the Red Sea, they found a daring plant that had made an unprecedented path of its own. They found an acacia tree that had moved soil and rock, through years of dry failure until its improbable success.

In my mind, in 1860, I see the men congratulate each other and gather around the root long enough to take a photograph with it. And then I picture them chopping it in half.

6

SCIENTISTS TAKE CARE of their own to the extent that they are able. When my undergraduate professors saw my sincere interest in their research laboratories, they advised me to continue on for a Ph.D. I applied for entrance to the most famous universities that I had ever heard of, giddy in the knowledge that if accepted, I'd get not only free tuition but also a stipend that would just cover rent and food for the duration of my enrollment. This is how Ph.D. training in science and engineering generally works—as long as your thesis also furthers the goals of a federally funded project, you are supported at a sort of academic subsistence level. The day after the University of Minnesota conferred upon me a bachelor's degree cum laude, I dumped off my winter clothes in a big pile at the Salvation Army on Lake Street, took Hiawatha Avenue south to Minneapolis–Saint Paul International Airport, and flew to San Francisco. After I got to Berkeley, I didn't so much meet Bill. It was more like I identified him.

During the summer of 1994, it became my responsibility to serve as the graduate student assistant instructor on what felt like an interminable field trip through the Central Valley of California. The average person cannot imagine himself staring at dirt for longer than the twenty seconds needed to pick up whatever object he just dropped, but this class was not for the average person. Each day for six entire weeks involved digging five to seven holes and stooping over them for hours, then camping out, and then doing it all over again at a different place. Every feature of every hole was subject to a complex taxonomy, and students were to become proficient in recording each tiny crack made by each plant root using the official rubric developed by the Natural Resources Conservation Service.

While examining a ditch of interest, the student employed the six-hundred-page *Keys to Soil Taxonomy*—a handy guide resembling a small phone book but much less interesting to read. Somewhere in Wichita (possibly), a committee of government agronomists has been perpetually enjoined to transcribe and reinterpret the *Keys* down through the ages as if it were an Aramaic text. The preface to the 1997 version of the *Keys* contains a moving passage describing the breakthroughs of the International Committee on Low Activity Clays that necessitated this new edition, emphasizing that it was written only for emergency use, given that the ongoing work of the International Committee on Aquic Moisture Regimes would likely make yet another overhaul unavoidable before 1999. But back in 1994 we were consigned to the 1983 version of the *Keys* and labored in childlike ignorance, little suspecting the bombshells soon to be dropped by the International Committee on Irrigation and Drainage.

We taught while crowded into a ditch with the ten-odd students who had worked with us to dig it out. The curriculum was designed to usher them into the secret world of the state agronomist, the civil servant, the park-service forester, and other practical land-management jobs. The grand finale of such soil-documentation exercises is the determination of "best use practice," for which one deems most suitable the construction of a "residential structure," a "commercial structure," or "infrastructure," after which one is goaded to "specify." By the fourth week, a septic tank seems far too posh an ornament for whatever hole your head is in, and so you resort to paving the mental landscape into one unending parking lot, which is how I suppose some portions of the United States got to be the way they are.

It took me about a week to notice that one of our undergraduate students—the one who looked like a young Johnny Cash and was perennially clad in jeans and a leather jacket even in 105-degree heat—always somehow ended up several meters away from the edge of the group, digging his own private hole. The main professor of the course was also my thesis advisor, and as his assistant, my role was largely behind-the-scenes. I floated from hole to hole, checking on the students' progress and answering any questions. I looked at the course roster and determined by process of elimination that the loner's name was Bill. I went over and interrupted his solitary work. "How are you doing? Do you have any questions or anything?"

Without looking up, Bill refused my help, saying, "Nah. I'm good." I stood there for a minute and then walked away and checked on another group, evaluated their progress, and answered some questions.

About thirty minutes later, I noticed that Bill was now digging a second hole, his first one having been carefully refilled and smoothed over at its top. I picked up his clipboard and saw that his soil evaluation had been completed meticulously and that he had also included his second-best answers in a separate column down the right side of the page. At the very top of his report, suitability for "infrastructure" was checked, and a specification of "juvenile detention center" had been added in careful handwriting.

I stood next to his hole. "Looking for gold?" I joked, trying to strike up a conversation.

"No. I just like to dig," he explained without stopping. "I used to live in a hole."

His matter-of-fact relation of this personal detail made me understand that he had meant it literally. "I also don't like for people to see the back of my head," he added.

Not taking the hint, I stood there and watched him dig for a while, and began to notice the uncommonly large amount of earth that he was moving with each shovelful and the implied strength that must have accompanied his wiry frame. I also noticed that he was digging with something that looked like an old harpoon flattened at one end—a sword beaten into a real plowshare. "Where did you get that shovel?" I asked, figuring it was from the pile of junk I had hauled out of the department's equipment locker, which had been located in the basement next to an old coal hull.

"It's mine," he said. "Don't judge it until you've dug a mile in its shoes."

"You mean you brought your own shovel from home?" I laughed in friendly surprise and delight.

"Hell yes," he affirmed. "I wasn't going to leave this thing unattended for six weeks."

"I like your thinking," I replied, seeing that I was clearly not needed. "Just let me know if you get stuck or have any questions." I started to leave, but hesitated when Bill looked up at last.

He sighed. "Actually, I do have a question. Why aren't those morons over there done already? This is like the hundredth hole we've looked at. How long does it take someone to learn to spot a fucking earthworm?"

I shook my head in corroboration and shrugged. "I guess their eye hath not seen, nor ear heard."

Bill looked at me for ten seconds. "What the hell is that supposed to mean?"

I shrugged again. "How should I know? It's from the Bible. You're not supposed to know what it means. Nobody does."

He looked at me suspiciously for a minute, but after he saw that I had nothing further to say, he relaxed and returned to digging. Later that evening, after the communally prepared dinner had been rationed out, I sat down at a picnic table across from him. Bill was wrestling with his undercooked chicken. "Wow," I remarked while examining my own plate. "I don't think I can eat this."

"I know. It's gross," he conceded. "But it's free, so I scarf down seconds each night."

"As a dog returneth to his vomit," I said, while making the sign of the cross in the air in front of me.

"Amen," he agreed with his mouth full, and toasted me with his 7Up can.

After this we began to casually seek each other out, and observing the larger action as a pair became a comfortable default position for both of us. We took to situating ourselves on one edge of the group—still part of it, but removed from the main activity. It seemed natural and easy that we should sit together much while talking little.

Each evening while I spent the hours reading, Bill sat and rubbed handfuls of dirt across the blade of his old Buck knife, rounding its edge past the dullness of a spatula. He explained to me in great detail how a knife is better for digging than a shovel when you are dealing with a very clayey soil.

"What's the book about?" he asked me one night.

I was reading a new biography of Jean Genet, with whom I had been fascinated since seeing a production of *The Screens* in Minneapolis in 1989. To me, Genet was the perfect representation of an organic writer, one who wrote purely and didn't labor to communicate, didn't expect recognition, and when recognition came didn't take it in. He was also untaught, which meant that his voice was absolutely original and not a subconscious imitation of hundreds of other books he'd read. I was obsessed with trying to figure out how Genet's early life had destined him for success while rendering him immune to it.

"It's about Jean Genet," I answered guardedly, knowing that I was revealing myself to be a bit of a nerd. Bill displayed no judgment and even some noncommittal interest. I ventured to explain. "He was a great writer of his generation—had a boundless and complex imagination—but even after he got famous, he just didn't realize it on some level."

I added some of the details that disturbed me most. "While he was growing up, he was incarcerated for one meaningless crime after another and so he developed an alternative vision of morality," I explained, surprised at how good it felt to be talking with someone about a book. Being outside in the fresh air while speculating on the motives of a dead author made me think of my family, from whom I had drifted far away, in every sense. I watched Bill scrape his knife through the dirt and remembered summer days in the garden with my mother.

"Genet worked as a prostitute and robbed his clients, and then used the time in jail to write books," I continued. "The weird thing is that even after he got wealthy, he would still go into stores and steal random stuff that he didn't need. Pablo Picasso personally bailed him out of jail once . . . It just doesn't make any sense," I concluded.

"It probably made perfect sense to him," Bill countered. "Everybody does all kinds of shit that they don't know why they do. They just know that they have to," he said, and I thought about that for a moment.

"Hey, you guys! Want a cold one?" We were interrupted by a good-natured offer from a drunkish student who was dangerously armed with a guitar. He was waving the sort of beer that one purchases for six dollars a case when miles from nowhere.

"No, I don't. That stuff you are drinking tastes like piss," Bill said.

I felt a need to soften Bill's statement and added, "Well, I don't really like beer, but that stuff does seem pretty awful."

"Jean Genet wouldn't have even stolen that shit," Bill hollered at him over his shoulder, and I smiled, knowing that the joke was ours alone.

The little group of students leaned in toward one another and said something private, and then began to titter in our direction. Bill and I looked at each other and rolled our eyes. It might have been the first time, but it certainly wouldn't be the last, that the people around us would misinterpret the nature of our connection.

During the next week we toured a working citrus orchard and were dumbfounded to learn how many different ways there were to mechanically shake the produce from a tree. We also toured the packing facility and saw rows of women standing along a conveyor belt pulling out large or oddly shaped spheres from a river of forest-green fruit that flowed down the line at a rate of ten per second. I am sure that we looked confused when our guide announced solemnly that these women were sorting lemons; it would have been easier to believe that the spheres were billiard balls, given the extremely hard knocking noise they made while bouncing down the conveyor belt.

Our guide loudly narrated our visit, gushing about how this factory was a great place to work, complete with on-site housing, and I thought about the weird little town that would result from such an arrangement. He ushered us into the plus-five-degree "ripening room," which was like a windowless train car packed floor-to-ceiling with the hard green fruit. The door would be sealed tonight, he told us, and the room would be flooded with ethylene gas, forcing these lemons to get off their asses and ripen in ten hours. Sure enough, the room next to us contained thousands of identically sized fruits, each sporting a peel so perfectly yellow that it could have been made of plastic.

After the tour was over, we milled about in the parking lot. "Good grief, talk about mind-numbing. I'll never complain about school again." Bill was referring to the lemon-sorting and was also jumping up and down in order to warm himself up after leaving the chilled rooms.

"Assembly lines depress the shit out of me. The town where I grew up had miles of them," I said, rubbing my hands and shuddering at the secondhand memory of my brother's gory third-grade field trip through the slaughterhouse. "Actually, they were more like disassembly lines."

"Did you ever work in the factory?" Bill asked.

"I was lucky, I went to college instead. I moved out of my parents' house when I turned seventeen." I spoke cautiously, modulating my urge to trust him.

"I moved out of my parents' house when I was twelve," Bill replied. "But not far, just into the yard."

I nodded, as if this was the most perfectly normal thing in the world. "Was that when you lived in a hole?"

"It was more of an underground fort. I put carpet and electricity in it and everything." He spoke offhandedly, but not without shy pride.

"Sounds cool," I said, "but I don't think I could sleep in a fort like that."

Bill shrugged. "I'm Armenian," he said. "We're most comfortable underground."

I didn't realize it at the time, but he was making a dark joke about his father, who as a child had been hidden in a well during the massacre that had killed the rest of his family. Later, I came to know that Bill lived pursued by the ghosts of his macabre ancestors, and it was they who continuously pressed him to build, plan, hoard, and—above all—survive.

"Where is Armenia? I don't even know," I asked.

"Most of it isn't anywhere," he answered. "That's kind of the problem."

I nodded, sensing the gravity of his words while not really understanding them.

Near the end of the trip, I approached my advisor as he readied the equipment for the next day's work. "Listen, we have to hire that Bill guy in the lab," I told him.

"You mean the weird dude who's always off by himself?" he asked.

"Yes. He's the smartest one in the class. We need him in the lab."

My advisor looked back at the tools he was sorting. "Uh-huh. And how do you know that?" he asked me.

"I don't know it," I said, "but I feel it."

As usual, my advisor relented. "Okay, go ahead, but you have to do the paperwork. I'm already way too overloaded, so he's your responsibility. You are the one who is going to keep him busy, got it?"

I nodded gratefully. I was newly excited about the future, but I didn't quite know why.

Three days later, when we finally rolled back into town after the trip's end, it was my job to drop the students off, finally bringing them home with their gear. Bill was the last to be delivered, and it was late at night as I pulled up to the BART station that he'd requested.

I mentioned the possibility of a job to him. "Hey, I don't know if you are interested, but I could set it up for you to work in the research laboratory where I work. For money and everything."

There was no immediate reaction to my statement. He looked down and after a moment he said gravely, "Okay."

"Okay, then," I agreed.

Bill continued to sit and stare at his feet while I waited for him to get out of the car and say goodbye. Presently he looked up and then out of the window for several more minutes while I wondered what could be keeping him.

Finally, Bill turned around and spoke to me: "Aren't we going to the lab?" he asked.

"Now? You want to go now?" I smiled at my new friend.

"I've got nowhere else to go," he said gamely, and then added, "and I've got my own shovel."

As happens at odd moments, a scene in a book that I had read came back to me and I thought again of Dickens, but this time *Great*

Expectations. I thought about Estella and Pip at the end of the story, and about how they stood exhausted but hopeful within a dusty garden, tasked with rebuilding a ruined house. I thought about how even though neither character knew what to do next, they could see no shadow of being parted.

7

THE FIRST REAL LEAF is a new idea. As soon as a seed is anchored, its priorities shift and it redirects all its energy toward stretching up. Its reserves have nearly run out and it desperately needs to capture light in order to fuel the process that keeps it alive. As the tiniest plant in the forest, it has to work harder than everything above it, all the while enduring a misery of shade.

Folded within the embryo are the cotyledons: two tiny ready-made leaflets, inflatable for temporary use. They are as small and insufficient as the spare tire that is not intended to take you any farther than the nearest gas station. Once expanded with sap, these barely green cotyledons start up photosynthesis like an old car on a bitter winter morning. Crudely designed, they limp the whole plant along until it can undertake the construction of a true leaf, a *real* leaf. Once the plant is ready for a real leaf, the temporary cotyledons wither and are shed; they look nothing like all the other leaves that the plant will grow from this point forward.

The first real leaf is built using only a vague genetic pattern with nearly endless room for improvisation. Close your eyes and think of the points on a holly leaf, the star of a maple leaf, a heart-shaped ivy leaf, a triangular fern frond, the fingery leaves of a palm. Consider that there can easily be a hundred thousand lobed leaves on a single oak tree and that no two of them are exactly the same; in fact, some are easily twice as big as others. Every oak leaf on Earth is a unique embellishment of a single rough and incomplete blueprint.

The leaves of the world comprise countless billion elaborations of a single, simple machine designed for one job only—a job upon which hinges humankind. Leaves make sugar. Plants are the only things in the universe that can make sugar out of nonliving inorganic matter. All the sugar that you have ever eaten was first made within a leaf. Without a constant supply of glucose to your brain, you will die. Period. Under duress, your liver can make glucose out of protein or fat—but that protein or fat was originally constructed from a plant sugar within some other animal. It's inescapable: at this very moment, within the synapses of your brain, leaves are fueling thoughts of leaves.

A leaf is a platter of pigment strung with vascular lace. Veins bring water from the soil to the leaf, where it is torn apart using light. The energy produced from this tearing apart of water is what glues sugars together after they are fixed from the air. A second set of veins transports the sugary sap out of the leaf, down to the roots, where it is sorted and packaged for either immediate use or longer-term storage.

A leaf grows by enlarging the string of cells located along a central vein; single cells on the perimeter eventually decide independently when to stop dividing. From this tip, smaller veins develop, eventually completing the network at the stem; thus the overall maturation proceeds from tip to base. Once the most daring portion of the leaf is complete, the plant puts horse before cart and begins to slide sugar back down and in, down to where it will be used to make more root, which will be used to bring up more water, which will be used to expand new leaves, which will pull back more sugar, and in this manner four hundred million years have passed.

Every once in a while a plant gets an idea to make a new leaf that changes everything. The spines on a cholla cactus are barbed like a fishhook, sharp and tough enough to puncture the leathery skin of a tortoise. They also reduce airflow across the cactus's surface, thereby reducing evaporation. They provide meager shade for the stem and a surface upon which to condense dew. The spines are actually the leaves of the cactus; the green portion is its swollen stem.

Probably within just the last ten million years, a plant had a new idea, and instead of spreading its leaf out, it shaped it into a spine, such as those we find today on the cholla cactus. It was this new idea that allowed a new kind of plant to grow preposterously large and live long in a dry place where it was also the only green thing around to eat for miles—an absurdly inconceivable success. One new idea allowed the plant to see a new world and draw sweetness out of a whole new sky.

8

ESTABLISHING YOURSELF as a scientist takes an awfully long time. The riskiest part is learning what a true scientist is and then taking the first shaky steps down that path, which will become a road, which will become a highway, which will maybe someday lead you home. A true scientist doesn't perform prescribed experiments; she develops her own and thus generates wholly new knowledge. This transition between doing what you're told and telling yourself what to do generally occurs midway through a dissertation. In many ways, it is the most difficult and terrifying thing that a student can do, and being unable or unwilling to do it is much of what weeds people out of Ph.D. programs.

On the day that I became a scientist, I stood in a laboratory and watched the sun come up. I was convinced that I had seen something extraordinary, and I was waiting for the new day to ripen into a reasonable hour at which I could make a telephone call. I wanted to tell someone what I had discovered, though I wasn't quite sure whom to call.

My Ph.D. thesis was built around the tree *Celtis occidentalis,* better known as the hackberry tree, which is found all over North America, common as vanilla ice cream and similarly uninspiring in appearance. Hackberry trees are indigenous to North America and were widely planted in cities in response to one of the innumerable casualties of the European conquest of the New World.

For hundreds of years, beetles—as well as people—have emigrated from Europe to the United States, arriving on ships and docking at ports across New England. In 1928 a hardy group of six-legged insect-pioneers left the Netherlands and homesteaded themselves under the bark of countless *Ulmus* trees. During the process, they also introduced a deadly fungus directly into each tree's bloodstream. The trees responded by shutting down their vascular systems vein by vein to try to limit infection and slowly starved themselves to death while unused nutrients pooled at their roots. Even today, Dutch elm disease continues to ravage the United States and Canada, and tens of thousands of trees succumb each year, pushing the overall death toll well into the millions.

In contrast, not much can kill a hackberry tree, which has been observed to endure both early frost and late drought with nary a loss of leaf. These thirty-foot-tall trees will never grow to be as majestic as their sixty-foot-tall elm predecessors; they ask only a moderate amount from their surroundings and earn our regard in proportion to their humility.

I was interested in *Celtis occidentalis* because of its prodigious fruit that superficially resembles a cranberry. If you pick one up and squeeze it, however, you'll find that the berry is as hard as a rock—mainly because it is a rock: just under its rosy skin is a shell harder than that of an oyster. This rocky structure serves as a mighty fortress for a seed that might have to pass through an animal gut, weather the rain and snow, and do battle with ruthless fungi for years prior to germination. The sediments of many archaeological digs are positively loaded with the stony remains of hackberry pits, as each tree produces millions of seeds during its lifetime. I hoped to develop an analysis of these fossil seed pits that would allow me to guess the average summer temperatures that occurred between the glaciations of the Midwest.

For at least the last four hundred thousand years, glaciers have expanded from the North Pole and then contracted periodically, regular as clockwork. During the short interim periods when the Great Plains have been ice-free, plants and animals migrated, interbred, and tested out new food sources and habitats. But just how hot were these interim summers—were they like the full-on sweltering summers of today, or were they just balmy enough to prevent snow from falling? If you've ever lived in the Midwest, you know that this distinction matters, but imagine how much more it matters to people living close to the land, with animal skins for shelter and a moving target for a food supply.

My thesis advisor and I could imagine all kinds of chemical reactions that would lock in the temperature of formation as each seed pit condensed out of the fruit sap. Our whole theory of temperature-setting-fruit-becoming-fossil was novel and also mysterious enough to keep easy answers out of reach. I devised a set of experiments intended to break the main question down into a series of smaller, discrete tasks. My first task was to figure out exactly how a hackberry seed formed and what it was made of.

To this end, I posted sentry around several living hackberry trees in Minnesota and South Dakota in order to compare cold with (relatively) warm environments. I planned to collect the fruit periodically over the course of a year. Back in the lab in California, I would cut hundreds of these fruits into paper-thin slices, and then describe and photograph them under the microscope.

When I looked through the microscope that magnified it by a factor of 350, the smooth surface of the hackberry pit resembled a honeycomb all stuffed full of something hard and crumbly. Using the concept of a peach pit as a place to start, I soaked several hackberry pits in an acid that I was sure could dissolve at least a bushel of peaches, and then examined what was left. The stuffing had dissolved out from within the honeycomb, leaving its lacy white scaffolding behind. When I placed the wee white structure in a vacuum and heated it to fifteen hundred degrees, carbon dioxide was released, which meant that there was something organic inside the white lattice—yet another puzzling layer.

The tree had grown a seed, spun a stringy net around it, coated the net in some kind of skeleton, and then stuffed the holes full of the same material that makes up a peach pit. By doing so, it protected the seed, giving it a better chance of sprouting and therefore growing into a tree, and perhaps begetting ninety generations of additional trees. If we were going to get any long-term climate data out of these fossil seed pits, this lacy white lattice was clearly a strongbox of information. And once I knew what this most basic part of the seed pit was made of, I'd be on my way.

Just as each type of rock forms differently, they each fall apart differently too. One way to distinguish among the different minerals that are the building blocks of rocks is to smash a sample thoroughly and expose it to x-rays. Each grain of salt in a saltshaker is a perfect cube when viewed up close. Grind one grain into a fine powder and you have shattered it into millions of tiny, perfect cubes. The inescapable cube shape of salt persists because the very atoms that comprise pure salt are bonded together in the shape of a square scaffold that outlines an endless number of cubes. Any break to this structure will occur along the planes of weakness that define these bonds, resulting in more cubes, all repeating the same atomic pattern right down to their smallest components.

Different minerals have different chemical formulas, reflecting differences in the number and type of atoms they contain and the way those atoms are bonded together. Such differences give rise to differences in shape that persist even in powder form. If one can figure out the tiny shapes present in a pinch of mineral powder—even the heterogeneous powder from an ugly, complicated rock—one can also determine its chemical formula.

But how to see the shape of these tiny crystals? After an ocean wave hits a lighthouse, a ripple bounces back across the ocean. The size and shape of this reflective ripple carry information about both the wave and the lighthouse. If we are anchored in a rowboat far away, we can distinguish a lighthouse with a square base from one that is rounded by the way the ripple hits us, provided that we have a very good idea of the size of the wave, its energy and timing, and the direction it has traveled. This is similar to how we work out the tiny shapes within mineral powder, using the ripples that bounce back, or diffract, from very small electromagnetic waves known as x-rays. A piece of film catches the ripples at their peaks, and their spacing and intensity allow us to reconstruct the shape off which they bounced.

In the fall of 1994, I asked permission for access to the x-ray diffraction laboratory that was situated across campus from my usual lab, and I was allotted some hours during which to use the x-ray source. I looked forward to my analyses with the same happy anticipation one brings to a baseball game: anything might happen, but it will probably take a long time.

After much deliberating, I had chosen to reserve the machine at night, but I wasn't sure that I had made the best choice. There was a creepy post-doc who worked in that lab, and I was uncomfortable with his surly demeanor. I'd seen how the slightest look or question could set this guy off on a rage, and he seemed particularly menacing toward the odd female who stumbled into his orbit. Thus I had a dilemma: If I came during the day, I'd be sure to see him, but there would be people around who might serve as human shields. At night, I'd likely have the place to myself, but on the odd chance that he did come in, I'd be an easy mark. In the end I signed up for a midnight shift and brought a three-quarter-inch ratcheting wrench along with me. I wasn't quite sure how I would actually defend myself with the tool if something happened, but just having the weight of it in my back pocket made me feel better.

When I got to the x-ray diffraction laboratory, I placed a glass sample slide onto the countertop, covered it in fixating epoxy, and sprinkled it with powder from the ground hackberry pit. I placed the slide into the diffraction machine and oriented everything carefully, and then activated the x-ray source. After lining up the strip chart, I said a silent prayer that its unobservable inkwell was full enough to last the entirety of the run, and then I settled in to watch and wait.

When a lab experiment just won't work, moving heaven and earth often won't make it work—and, similarly, there are some experiments that you just can't screw up even if you try. The readout from the x-ray displayed one clear, unequivocal peak at exactly the same angle of diffraction each time I replicated the measurement.

The long, low, broad swoop of ink was totally unlike the stiff, jerky spikes that my advisor and I thought we might see, and it clearly indicated that my mineral was an opal. I stood and stared at the readout, knowing that there was no way I had—or anybody could have—possibly misinterpreted the result. It was opal and this was something I knew, something I could draw a circle around and testify to as being true. While looking at the graph, I thought about how I now knew something for certain that only an hour ago had been an absolute unknown, and I slowly began to appreciate how my life had just changed.

I was the only person in an infinite exploding universe who knew that this powder was made of opal. In a wide, wide world, full of unimaginable numbers of people, I was—in addition to being small and insufficient—special. I was not only a quirky bundle of genes, but I was also unique existentially, because of the tiny detail that I knew about Creation, because of what I had seen and then understood. Until I phoned someone, the concrete knowledge that opal was the mineral that fortified each seed on each hackberry tree was mine alone. Whether or not this was something worth knowing seemed another problem for another day. I stood and absorbed this revelation as my life turned a page, and my first scientific discovery shone, as even the cheapest plastic toy does when it is new.

I didn't want to touch anything, because I was just a visitor. So I stood and looked out the window, waiting for the sun to come up, and eventually a few tears ran down my face. I didn't know if I was crying because I was nobody's wife or mother—or because I felt like nobody's daughter—or because of the beauty of that single perfect line on the readout, which I could forever point to as *my* opal.

I had worked and waited for this day. In solving this mystery I had also proved something, at least to myself, and I finally knew what real research would feel like. But as satisfying as it was, it still stands out as one of the loneliest moments of my life. On some deep level, the realization that I could do good science was accompanied by the knowledge that I had formally and terminally missed my chance to become like any of the women that I had ever known.

In the years to come, I would create a new sort of normal for myself within my own laboratory. I would have a brother closer than any of my siblings, someone I could call any hour of the day or night and gossip with more shamelessly than I ever had with my girlfriends. Together, we would devote ourselves to exposing the absurdity of our endeavors and continuously remind each other of particularly ridiculous examples. I would nurture a new generation of students, some of whom were just hungry for attention, and a very few who would live up to the potential that I saw in them. But on that night, I wiped my face with my hands, embarrassed to be weeping over something that most people would see as either trivial or profoundly dull. I stared out the window and saw the first light of the day spilling its glow out upon the campus. I wondered who else in the world was having such an exquisite dawn.

I knew that before noon I would be told that my discovery was not special. An older and wiser scientist would tell me that, in fact, what I had seen was something that he himself might have assumed. While he explained that my observation wasn't a true revelation, only a confirmation of what should have been an obvious guess, I listened politely. It didn't matter what he said. Nothing could alter the overwhelming sweetness of briefly holding a small secret that the universe had earmarked just for me. I knew instinctively that if I was worthy of a small secret, I might someday be worthy of a big one.

By the time that the sunrise had burned through the Bay Area fog, I felt lifted out of my maudlin mood as well. I walked back to the building where I usually worked in order to start my day. The chilly air smelled of eucalyptus in a way that will always remind me of Berkeley, though the campus was quiet as death. I let myself into the lab and was surprised to find that the lights were on. I then saw Bill, who was sitting on an old lawn chair in the middle of the room and staring at a blank wall while listening to the static of talk radio on his little transistor.

"Hey, I found this chair in the Dumpster behind McDonald's," he told me as I walked in. "It seems to work." He examined it with satisfaction while still sitting upon it.

I felt deeply happy to see him. I had anticipated at least three more lonely hours of waiting for someone to talk to.

"I like it," I told him. "Can anybody sit in it?"

"Not today," he said. "Maybe tomorrow." He considered and then added, "But maybe not."

I stood and thought about how every single thing that came out of this guy's mouth was just a little on the weird side.

Against my Scandinavian instincts, I decided to tell Bill about the most important thing that I had ever done. "Hey, have you ever seen an x-ray of an opal?" I asked, holding up my paper readout.

Bill reached for his radio and silenced it by pulling out its nine-volt battery—the on-off switch had stopped working long ago. After he finished, he looked up at me. "I knew I was sitting here waiting for something," he told me. "Turns out it was that."

* * *

After I discovered that the hackberry pits contained opal, my next goal was to discern a way to back-calculate the temperature that governed its formation within the seed. While the scaffolding of the hackberry shell was indeed made of opal, the crumbly stuffing was made from a carbonate mineral called aragonite—the exact same mineral found in a snail's shell. Pure aragonite is easy to precipitate in the laboratory; one just mixes two supersaturated fluids, and the crystals rain out of the clear mixture like mist condensing within a cloud. The isotope chemistry of the crystals is strictly controlled by temperature, which means that by measuring the oxygen isotope signature of a single crystal, we could predict the exact temperature at which the solutions were mixed. I could make this work in the lab one hundred times out of a hundred. It was foolproof. My next task was to show that it also worked within a tree, that the same process was happening inside the fruit, where aragonite crystals formed as tree-sap solutions mixed.

My advising professor had pitched this idea as a fifteen-page grant proposal to the National Science Foundation, the peer-reviewers had liked it, and we were recommended for funding. And so, in the spring of 1995, I was headed back to the Midwest to look for the perfect trees to study. I decided upon three full-grown hackberry trees that I found growing on the banks of the South Platte River near Sterling, Colorado, less than a day's drive from a couch where I was always welcome. Under what felt like the biggest, bluest sky in the world I calculated how the composition of the river, taken with the composition of that summer's fruit, would allow me to solve for the average temperature of the season. Confident of success, I corded off the trees and began to monitor them like an expectant father—delighted in anticipation of the gift, but tangential to the proceedings. I also became similarly bewildered during the thick of things, because during that particular summer none of the hackberry trees at or near the site flowered or bore fruit.

Nothing in the world exposes human helplessness and folly quite like a tree that will not bloom. Unaccustomed to people—let alone things—that wouldn't eventually do what I wanted them to do, I took it hard. I analyzed the situation with my only friend in Logan County, Colorado: a guy named Buck who worked behind the counter in a liquor store at the highway crossing. I had gone into the store more desperate for air-conditioning than for beer, truth be told, but after Buck carded me he grudgingly admitted that I was "holding up pretty good for an old lady," and I took it as an invitation to hang around. As the summer wore on, Buck was increasingly bemused that he was having more luck with his scratch-offs than I was with my trees, but he refrained from rubbing my nose in the irony of my prior lectures on lottery statistics.

Buck had grown up on a ranch nearby, and so I vaguely felt he was a party to the whole fruit debacle, or at least that he should be answerable to it. "But *why* didn't they bloom? Why *this* year?" I urged the question on Buck. I had pored over the local climate records and found nothing conspicuous in the weather.

"It just happens sometimes. Somebody around here could have told you that," he said, dispensing the grim pity that is rarely to be had from cowboys.

I was convinced that the trees were giving me a sign and that my future career was unraveling. I was panicking, picturing myself on the assembly line, trimming the jowls off of dismembered hog heads, one after the other after the other, for six hours a day, just as the mother of my childhood friend had done for nearly twenty years. "That's not good enough," I answered. "There has to be a reason."

"Trees don't *have* a reason, they just do it, that's all," Buck snapped. "In fact, they don't *do* anything, they're just trees, they just *are*. Shit, they're not *alive*, not like you and me." He had finally gotten fed up, and something about me and my questions was irritating him.

"Ke-rist-on-a-crutch," he added in frustration, "they're just *trees.*"

I left the shop and never went back.

I returned to California in failure. "Well, if I had a car that I thought could make it over the Concord Bridge I'd say let's go set one of those trees on fire," Bill said as he concentrated the crumbs from the bottom of his Lay's potato chip bag using one of the lab's funnels. "We'll let the others watch that one burn for a while and then ask them if they don't feel a little more inclined to bloom."

Bill had become a fixture in my advisor's laboratory. He appeared sometime around 4:00 p.m. each day and then stayed for eight or ten hours, as his spirit and our needs moved him. He couldn't see how the fact that he was given pay for only ten hours a week was relevant, and he was surprisingly content to listen to me talk obsessively about my trees for hours each night while we worked. Before my last trip to Colorado, Bill had urged me to take a BB gun and indulge in a couple of afternoons of shooting at leaves and branches.

I declined. "Not that I'm an arborist or anything, but I don't think it will help."

"It will make you feel better," he said emphatically. "Trust me."

That whole summer in Colorado was a data-gathering bust, but it taught me the most important thing I know about science: that experiments are not about getting the world to do what you want it to do. While tending to my wounds that fall, I shaped a new and better goal out of the debris of the disaster. I would study plants in a new way—not from the outside, but from the inside. I would figure out why they did what they did and try to understand their logic, which must serve me better than simply defaulting to my own, I decided.

Every species on Earth—past or present, from the single-celled microbe to the biggest dinosaur, daisies, trees, *people*—must accomplish the same five things in order to persist: grow, reproduce, rebuild, store resources, and defend itself. At twenty-five, I could already see that my own reproduction was going to be complicated, were it ever to take place at all. It seemed outrageous to hope that fertility, resources, time, desire, and love could all come together in the right way, and yet most women did eventually walk that path.

While in Colorado, I'd been so focused upon what the trees weren't doing that I hadn't made any observations of what they were doing. Flowering and fruiting must have taken a backseat to something else that summer, something that I had failed to notice. The trees were *always doing something:* when I kept this fact placed firmly in front of me, I got closer to making sense of the problem.

A new mind-set became imperative: perhaps I could learn to see the world as plants do, put myself in their place, and puzzle out how they work. As a terminal outsider to their world, how close could I come to getting inside? I tried to visualize a new environmental science that was not based on the world that we wanted with plants in it, but instead based on a vision of the plants' world with us in it. I thought of the different labs that I had worked in and the wonderful machines, chemicals, and microscopes that gave me so much happiness . . . What kind of hard science could I bring to bear on this weird quest?

The perversity of such an approach was seductive; what was there to stop me, aside from my own fear of being "unscientific"? I knew that if I told people I was studying "what it's like to be a plant," some would dismiss me as a joke, but perhaps others might sign on just for the adventure. Maybe hard work could stabilize scientifically shaky ground. I didn't know for sure, but I felt the first delicious twinges of what would be my life's enduring thrill. It was a new idea, my first real leaf. Just like every other audacious seedling in the world, I would make it up as I went along.

9

EVERY PLANT CAN BE SEPARATED into three components: leaf, stem, and root. Every stem functions the same way: as a bundle of bound straws, bales of microscopic conduits that carry soilwater up out of the roots and sugarwater down out of the leaves. Trees are a unique type of plant because their stems can be more than one hundred yards long and are made of this amazing substance that we call wood.

Wood is strong, light, flexible, nontoxic, and weather-resistant; thousands of years of human civilization have yet to produce a better multipurpose building material. Inch for inch, a wooden beam is as strong as one made from cast iron but is ten times more flexible and only one-tenth as heavy. Even in this age of high-tech man-made objects, our preferred construction material for housing remains lumber hewn from trees. In the United States alone, the total length of the wooden planks used during the last twenty years was more than enough to build a footbridge from the planet Earth to the planet Mars.

People slice up tree trunks, nail the pieces together into boxy shapes, and then go inside to sleep. Trees use the wood in their trunks for a different purpose—namely, they use it to fight with other plants. From dandelions to daffodils, from ferns to figs, from potatoes to pine trees—every plant growing on land is striving toward two prizes: light, which comes from above, and water, which comes from below. Any contest between two plants can be decided in one move, when the winner simultaneously reaches higher and digs deeper than the loser. Consider the tremendous advantage that wood confers to one of the contestants during such a battle: armed with a stiff-yet-flexible, strong-yet-light prop that separates—and connects—leaves and roots, trees have dominated the tournament for more than four hundred million years.

Wood is a static, utilitarian compound, constructed once and left to stand as inert tissue forevermore. From the tree's center (or "heartwood") radiates a network of ray cells that bring cool xylem and sweet phloem to the cambium layer on the periphery. The cambium layer manufactures the living sheath that rests just below the bark. A tree grows by producing one new sheath after another. When a sheath is outgrown, its woody skeleton is left behind, progressively forming the rings that we can see in cross section after a tree is felled.

A tree's wood is also its memoir: we can count the rings to learn the tree's age, for every season of growth requires a new sheath from the cambium. There's a lot of additional information written into tree rings, but it is coded within a language that scientists don't speak fluently—yet. An unusually thick ring could signify a good year, with lots of growth, or it could just be the product of adolescence, a random spurt of growth hormones cued by an influx of unfamiliar pollen from a distant source. A ring that is thick on one side of the tree but thin on the other tells the story of a fallen branch. When a branch is lost, it upsets the balance of the tree, triggering cells within the trunk to reinforce the side that must now support the newly uneven burden of the crown.

For trees, losing limbs is the rule, not the exception. The vast majority of the branches that any tree produces are severed before they become large, usually by external forces like wind, lightning, or just plain old gravity. Misfortunes that cannot be prevented must be endured, and trees possess a ready strategy. Within a year after the loss, the cambium will cast a healthy new sheath fully over the broken base of what used to be the branch, and then layer upon it year after year until no scar is visible at the surface.

In the city of Honolulu, just where Manoa Road crosses Oahu Avenue, there stands a gigantic monkeypod tree (*Pithecellobium saman*). The trunk of this tree is fifty feet tall and its branches form a giant arch that spans clear across the busy intersection. Wild orchids grow on the branches: they sit in companionable bunches shaped like pineapple tops and their naked roots dangle down beneath them. Feral parrots hop from one orchid to another, flapping their lemon-lime wings and squawking abuse at the pedestrians below.

The monkeypod, like many trees of the tropics, lives in eternal bloom: great globs of silky, threadlike pink and yellow petals rain down on the tourists who pause to take a picture of the tree as they make their way up the valley to visit the famous Manoa waterfalls. On coffee tables all over the world, you can find photo albums containing photographs of the monkeypod tree at Manoa Road and Oahu Avenue, thousands of worm's-eye views of its magnificent eight-thousand-square-foot canopy, woven through with flowers.

From the tourists' perspective, this tree has achieved its perfect form: they do not see a tree that is less than it might have been, or one that was forced to grow a different way after its limbs were torn. If the monkeypod tree at Manoa Road and Oahu Avenue were to be cut down, we could count the knots and see the buried scars of the hundreds of branches that it has lost during the last century of its life. But as of today the tree stands, and while it is standing, we see only the branches that did grow and do not miss the ones that were lost.

Every piece of wood in your house—from the windowsills to the furniture to the rafters—was once part of a living being, thriving in the open and pulsing with sap. If you look at these wooden objects across the grain, you might be able to trace out the boundaries of a couple of rings. The delicate shape of those lines tells you the story of a couple of years. If you know how to listen, each ring describes how the rain fell and the wind blew and the sun appeared every day at dawn.

10

THE REMAINDER OF 1995 went by quickly. Once I had passed the prerequisite and grueling three-hour oral exam that qualified me to write a dissertation, nothing was left but to write it. I did that quickly, indulging myself in long writing jags and typing with the TV on in order to supply the noise that I needed in order to focus past my loneliness. Soon after my thesis was written, I graduated. The four years that I had spent on my Ph.D. seemed to have passed in the blink of an eye. Knowing that I'd have to be at least twice as proactive and strategic as my male counterparts, I had started applying for professor jobs during my third year and had successfully secured an offer at a quickly growing state university: Georgia Tech. The next phase of my career was coming into focus, or so everyone kept telling me.

In May 1996, Bill was awarded his bachelor's degree at the same ostentatious ceremony where I was awarded my doctorate. Neither one of us had any family in attendance, and so we found ourselves awkwardly shifting about on the sidelines while the other graduates were hugged and photographed, everyone beaming over their diplomas. After an hour of this, we agreed that no free glass of champagne was worth the torture, and we walked back to the lab. We took off our graduation robes, wadded them up, and threw them in a corner. Once we got our lab coats on, everything felt more normal. The night was still young: it was barely nine o'clock and prime working hours hadn't even started yet.

We decided to spend the night blowing glass, which was our favorite late-night diversion. My goal was to seal a tiny amount of perfectly pure carbon dioxide gas into each of about thirty glass tubes. I would need them for reference when running the mass spectrometer—each tube would provide a watermark of known value against which I could compare my unknown samples. Making these "references" was a time-consuming task that needed to be repeated approximately every ten days, and like a lot of lab work that happens in the background, it wasn't very interesting, but at the same time it was key that I do it carefully and without error.

Bill sat nearby and performed the first step in the process by melting one end of a length of glass tube. In order to melt the glass, he was using a torch set to a small flame, powered by acetylene fuel and boosted by a stream of pure oxygen gas. It was like an entire barbeque grill on steroids, all blasting out of a single tiny opening—that was pointed away from his face, of course. The flame that emerges from this kind of torch is so bright that it will damage the human eye if you look directly at it, and so we were both wearing dark safety goggles.

Glass is hard and brittle at room temperature, but it softens into luminescent toffee when heated to a few hundred degrees. Melted glass is hot enough to ignite paper or wood on contact. A drop of molten glass readily burns down through the skin if you spill it onto your arm, stopping only when cooled by a bone-deep rush of blood. University policy probably dictated that I shouldn't have been giving an undergraduate such a dangerous and advanced duty, but Bill had easily learned all the lesser jobs that I'd shown him, then proceeded to fix everything that was broken, and had finally gone on to perform preventative maintenance—all on his own initiative. He was simply running out of things to do, and I couldn't justify not letting him advance to more important tasks, so I taught him the basics of how to blow glass.

While we worked during that night, I looked forward across my life and saw a future where I was making reference tubes weekly forevermore, shriveling up and turning gray while watching the dancing needle of a gauge, just like the one in front of me. The thought was depressing and comforting at the same time. I knew only one thing for certain: that I couldn't imagine any other future for myself.

Snapping out of my daydream, I looked past the liquid nitrogen trap and toward the gauge. Its needle read flat, indicating that there was no gas left in the line; it had all condensed into my tube and was frozen inside the trap. I sealed the glass tube by melting it shut and then set it down such that the molten side would cool slowly while its frozen contents thawed.

I looked over my shoulder to see Bill engrossed in producing tubes. "How about some radio?" I asked him conspiratorially, offering to break up the monotony with the rare treat of superfluous noise. As a rule, music is forbidden in the laboratory, especially during dangerous and painstaking work. We'd been trained to understand that you can't afford to have any part of your brain distracted from what you are doing when each move you make is critical to both safety and success.

"Yeah, sure," he agreed. "Anything except that NPR shit. I don't want to get all worked up over the plight of fishermen in some place that I can't even find on the map. I got my own problems right here."

I thought I understood this remark, but I kept my mouth shut and did not comment. I had recently dropped Bill off in front of a dingy apartment complex on the border of what was known to be a crime-ridden neighborhood in Oakland, so while I was pretty sure that he wasn't actually homeless, I still suspected that it wasn't a good scene. For all the time that we spent together, Bill had mostly remained a mystery to me. I had been around him enough to know that he didn't do drugs, skip class, or litter on the street—incongruously enough, considering his disaffected comportment—but I didn't know anything beyond that.

I took off my safety goggles, bent down behind the stereo, and began to search the frequencies, looking for an AM radio talk show that might entertain us for a while. The broken knob was not well engaged with the tuning mechanism, and so I had to fiddle with the dial in order to get it to move at all. The last thing I remember hearing clearly was an unbelievably loud, sharp pop, as if someone had set a firecracker off inside my head. After that I didn't hear anything at all for about five minutes. Nothing. Not my own breathing, not the buzzing of the building's airflow system, not the whoosh of my blood pulsing through my head. Nothing.

Terrified, I stood up and saw that the side of the lab where I had been working was now showered in splintered glass. I careened my head around and saw then that I was alone. Bill wasn't where he had been sitting. I panicked and yelled his name. When I didn't hear my own yell, my panic increased. Then I saw Bill's head slowly rising over the counter, looking at me with his eyes wide as saucers. When he'd heard what sounded like a gunshot at close range, he dove under a desk and remained crouched there until he heard me shouting his name.

All at once it dawned on me what had gone wrong. I had condensed more carbon dioxide into that glass tube than I intended to. I had left it a minute too long while daydreaming and far more gas had been trapped than the tube could hold. After I sealed the tube, the frozen gas had warmed up, expanding quickly and exploding like a pipe bomb. Moreover, it had exploded into Bill's stockpile of other glass tubes, shattering several days' worth of work and sending glass splinters flying throughout the room.

Embedded in the back of the radio were hundreds of tiny shards of glass, and some not so tiny. The stereo had miraculously shielded my face from the explosion; had I not been occupied in finding a station, the shower of glass would no doubt have hit me in my eyes. I was struck by the irrational fear that everything in the room might explode and I looked around wildly, finally understanding that we were now safe, if only because all our glass was already broken. My hearing started to slowly come back and with it came a mighty earache, which made every whispering sound burn the inside of my head as if my ear canals were raw and bleeding.

I cannot do this, I thought, and then: *What the hell did I think I was doing here?* I had fucked up. This was bad.

Bill turned off the torches and then walked systematically around the room, unplugging everything. I stood there wondering what to do. I felt as if my whole world had exploded along with the tubes. *Scientists don't do things like this. Fuck-ups do things like this,* I thought. I couldn't even look Bill in the eye.

"Hey, can I take a cigarette break?" Bill asked at length with a surprising calm that served to make the whole thing even more unreal.

I nodded, wincing. My ears hurt like crazy.

Bill shuffled through the glass splinters that lay like hailstones all around us, toward the door. When he got there, he stopped and turned around. "You coming?" he asked me.

"I don't smoke," I answered miserably.

Bill jerked his head toward the hall. "That's okay," he told me. "I'll teach you."

We went outside, walked a few blocks down Telegraph Avenue, and then sat down on the curb. Bill lit a cigarette and we shivered in our T-shirts, queasy in the chilly Northern California night. The usual nocturnal characters who meandered around Berkeley were out and we watched them walk by, some talking frantically to themselves.

I drew my knees up to my chest and began to chew on the backs of my hands. It was a habit that I worked diligently to hide from other people. In the lab I could usually just wear gloves, but at that moment a great anxiety was overtaking me. I worked the knuckles of my right hand with my teeth until I felt the thin scabs open, and the taste of blood and the feeling of tearing skin began to calm me in a way that nothing else ever did. I ground my teeth into the raw skin between my knuckles, teething on my bones, sucking desperately for comfort. In a few short months I would be a professor, but that night I was pretty sure I couldn't do anything.

Bill took a drag. "We used to have a dog that chewed her paws," he reflected.

"I know it's gross," I said, flooded with shame. I doubled over my hands and pushed them into my stomach in an effort to keep them out of my mouth.

"No," he said, "she was a great dog. We didn't give a damn." He continued, "When you have a dog that good, you let it do whatever it wants." I rested my head on my knees with my eyes shut tight. We sat in silence while Bill smoked.

At length we walked back to the lab and swept up the glass, working carefully to hide all traces of what had happened. I was glad that it was the middle of the night, but felt guilty when it became clear that I was going to get away with such a serious mistake.

"Hey, what are you going to do next year, do you know?" I asked Bill as we swept. I had been unsurprised to learn that Bill had earned honors for his bachelor's degree in soil science, and I naturally assumed that he had some job lined up, as our department was known for its ability to place its graduates.

"My plan," explained Bill matter-of-factly, "is to dig another hole in my parents' yard and move into it." I nodded in acknowledgment. "And smoke," he said, "until I run out of cigarettes." I nodded again. "Then I guess I'll probably chew on my hands," he added with a shrug.

I hesitated, and then I took the plunge. "Listen, do you want to move to Atlanta and help me build a lab?" I asked, and then added, "I can pay you. Well, I'm pretty sure I can, anyway."

He thought for a while. "Can we bring that radio?" he asked, pointing at the shredded plastic stereo we were about to ditch in the Dumpster out back.

"Yes," I said. "We'll get a bunch of them."

* * *

Two months later we loaded up all of our belongings—which fit easily into my pickup truck—and drove to Southern California, where I dropped Bill off with his family at his childhood home. We had agreed that I would move first, in time for the fall semester to start at Georgia Tech, and that he would join me a few months later.

Bill's parents were extremely warm and friendly, generous and hospitable hosts, who from that first meeting treated me as a long-lost daughter. When I met him, Bill's dad was about eighty years old and had fascinating stories to tell, having worked his entire career as an independent filmmaker, documenting firsthand accounts of the Armenian genocide from which his family had fled when he was a boy. Supported in piecemeal fashion by the National Endowment for the Arts, Bill's whole family had worked to make these films happen, and Bill and his brothers had grown up serving as the film crew while hard-traveling through Syria. In their family home near Hollywood they edited footage in the studio and tended a huge garden; his dad could make anything grow and his mom insisted that I eat only the oranges from their very best tree.

On the final night of my visit, I lay on the bed in Bill's sister's bedroom, staring up at the ceiling and thinking about my future. The next morning, I would drive up to Barstow, merge onto Interstate 40, and leave California for good. It wasn't the first time that I would walk away from everything I knew and everything to which I'd become attached, knowing that I could never go back. It was the same way that I'd left home for college and then left college for graduate school: everyone but me was sure that I was ready to go. It was, however, the first time that I'd have a for-sure friend at the place where I was going, and I knew enough to thank God for that.

* * *

On August 1, 1996, I officially became an assistant professor at Georgia Tech, and I was expected to look and act like one, even though I was only twenty-six years old and I had no concrete idea of how one does either of those things. There were many days when I'd spend six hours preparing a one-hour classroom lecture. Afterward, I'd reward myself by sitting in my office choosing and ordering chemicals and equipment, feeling like a giddy bride picking out her gift registry. When my purchases arrived, I piled them up in the basement, and soon they formed a mountain of cardboard boxes. The central mailroom had scrawled the word "JAHREN" on each box when it was received, and it struck me as beautiful when I leaned against the wall and stared at the giant tower bearing my name lettered in twenty different hands. Bill was scheduled to arrive in January and then we would begin to set everything right, to make real the dreamscape that we had so often described to each other back in California. I didn't want to open any of the packages until we could do it together—but I was like a little kid waiting for Christmas morning. I would pick up a box, shake it, try to guess exactly what was inside, start to open it, restrain myself, and then return it to the pile.

I taught geology to freshmen and geochemistry to juniors, and it was far more work than I had reckoned it would be. During that first semester, I think I made more mistakes on the homework than the students did. Eventually, I embraced the persona of the amiable, forgiving professor who was eager to give everyone an A. It suited me better than trying to be a hard-ass, as I wasn't much older than most of the undergrads, and I was younger than many of the graduate students. For my part, I had never liked lecture courses anyway—everything important that I had learned had come to me from working with my hands.

Nevertheless, I dutifully fulfilled my lecturing obligations. I wrote equations on the chalkboard, assigned and graded homework, held office hours, and gave final exams, but I mostly focused on the coming New Year, when Bill and I would start building my first-ever all-my-own lab.

On the day that Bill flew in, I drove to the Atlanta airport an hour early and then stood in the baggage area mesmerized by the circling carousels. Suddenly I heard a familiar voice: "Hey, Hope, over here." I turned and saw Bill standing two carousels over. He was burdened with four heavy suitcases of the old hard-sided kind without rollers or straps.

"Oh, hi."

I had been at the wrong baggage claim. Confused, I looked around me. I didn't remember checking the carousel number. I also didn't remember parking the car, and yet there I was, holding a ticket to the parking structure with the location "C2" written on it in my own handwriting. This sort of thing was happening to me frequently: snippets of time were lost here and there, and even as I compensated in order to hide it, it kept getting worse. I had even gone to see a doctor about it, who had examined me for all of forty-five seconds, told me I was working too hard, and then wrote me a refillable prescription for a mild sedative.

"You look different," said Bill.

He was right. I wasn't sleeping much and I had lost quite a bit of weight. I had always been high-strung, but this felt like something different.

"I have anxiety. It's my new thing," I explained, opening my eyes very wide. "It affects more than twenty-five million Americans," I said, quoting the pamphlet that the doctor had given me.

"Okay." Bill looked around and added, "So this is Atlanta. Jesus, what are we even doing here?"

"It's our last best hope for peace!" I quoted the opening lines of

Babylon 5 in a deep science-fiction-narrator voice. I laughed at my joke, but Bill didn't.

We walked across the skyway into the parking garage and found my car. Bill stuffed his belongings into the back and then slid into the passenger seat. "I've never been this far east," he revealed. "They do sell cigarettes in these parts, don't they?"

I handed over the unopened pack of Marlboro Lights that had been in my purse for months. "Sorry I haven't kept up with my practicing. But I am getting pretty good with these things." I showed him my prescription bottle of lorazepam, shaking it like a rattle.

"To each his own," muttered Bill. He lit a cigarette, rolled down the window, and threw out the spent match. I inhaled his secondhand smoke and relaxed into its familiar smell. Bill was delighted to find that winter means very little in the South, and we drove with the windows down and without our seat belts on, making our way inside the beltway and toward the rising skyline of Atlanta. I felt the deep and simple happiness that comes from not being alone.

After driving for a few more minutes, I realized that I didn't know where I was supposed to be taking Bill. That night two and a half years earlier when we had returned from the field course and he was the last to be dropped off came to mind.

I offered, "You know you're welcome to crash on my couch for a while, until you rent a place of your own."

"No thanks. Just dump me downtown later and I'll figure it out," he said. "Right now, I want to see the new lab."

"Okeydokey," I agreed. "Let's go."

We drove to the university and I parked outside of our building, which was known as "Old Civil Engineering" even though the engineering department had long since relocated to better digs. I escorted Bill down the stairwell, into the basement, and to the room that would serve as our laboratory. I could barely contain my excitement as I turned the key and opened the door.

Once I got the door open, however, it occurred to me that I had nothing much to show him. It was a windowless room, about six hundred square feet, and when I looked at it through a visitor's eyes, I became aware of how little it resembled the glittering high-tech space that I had described to Bill during our many daydream sessions in California.

I looked around the dingy little room that had been battered and then abandoned. The drywall was pockmarked and ripped in places. Light switches spilled out of gashes in the walls and dangled freely from their hasty wiring. The wiring of a splayed power outlet was unraveled near our feet. A layer of rusty mildew covered everything, including the fluorescent light tubes that flickered above our heads. All around the perimeter, where there should have been wainscoting, there was instead a long, dry smear of something that might once have been glue. The area near the chemical hood stank like rancid formaldehyde, which was a bad sign given that the sole purpose of a working chemical hood is to prevent one from breathing, and thus smelling, chemicals.

Looking at Bill, I felt a sudden urge to apologize for the inadequacy of it all. The tour was just getting started, and I was already ashamed that this was all I had to offer to someone who had moved so far away from home at my invitation. It was nothing like the lab where we had worked in Berkeley, and it clearly never could be.

Bill removed his coat and threw it in a corner. He took a deep breath, ran both hands through the full length of his hair, and turned around slowly, counting the electrical outlets. He caught sight of the transformer and power conditioner that had been haphazardly installed in one corner of the room, complete with a bright-red emergency power cutoff. He pointed at it and said, "Oh, this is great. This will give us a stable two-hundred-twenty-volt supply. Exactly right for the mass spectrometer. Just completely perfect," he added for emphasis.

It was what it was: the first labspace to which we were the sole possessors of the key. It may have been a tiny hellhole, but it was *ours*. I marveled that Bill could see that gutted room not in comparison with what we had always planned it would be, but for what it was, and for what hard work might make it into. Despite the large difference between past dreams and present reality, he was ready to love our new life. I made up my mind that I would try to love it too.

11

IT'S RARE, but a single tree can be in two places at once. Two such trees can exist up to one mile apart and yet still be the same organism. These trees are more similar than identical twins. In fact, they are identical without qualification right down to each single gene. If you cut both trees down and count the rings you will see that one of them is much younger than the other. When you sequence their DNA, however, you will find no differences. This is because they used to be parts of the same tree.

It is easy to become besotted with a willow. The Rapunzel of the plant world, this tree appears as a graceful princess bowed down by her lush tresses, waiting on the riverbank for someone just like you to come along and keep her company. Don't be fooled into thinking that your fairy-tale willow is special, however. Chances are that she is not. If you walk upstream, it is likely that you will find another willow tree. It is also likely that this tree will be precisely the same willow as your dear willow, standing in a different pose, with a different height and girth, and having perhaps seduced dozens of other princes over the years.

A willow tree is far more like Cinderella than it is like Rapunzel in that her lot in life involves working harder than her sisters. There's a famous study in which scientists compared the growth rates of a group of trees for a year. The hickory and buckeye were fast out of the gate, but then stopped growing after just a few weeks. The poplar made a good show and grew for four full months. But it was the willow that quietly outpaced all the others, continuing to grow for a full six months through the shortening days of autumn and right up to winter's iron gates. The willow trees in the study grew an average of four feet by the end—almost double the growth of their nearest competitor.

Light equals life for a plant. As a tree grows, its lower branches become obsolete, too shaded by the newer ones above to be of any further use. A willow tree loads these used branches with reserves, fattens and strengthens them and then dehydrates their base such that they snap off cleanly and fall into the river. Carried away on the water, one out of millions of these sticks will wash up onto a bank and replant itself, and before long that very same tree is now growing elsewhere. What was once a twig will be forced to function as a trunk, stranded under conditions it had never considered. Every willow tree features more than ten thousand such snap-off points; it sheds 10 percent of its branches in this way every single year. Over the decades one—maybe two—of these will successfully take root downriver and grow into a genetically identical doppelgänger.

The oldest surviving family of plants on Earth is the *Equisetum*—the horsetail. The fifteen or so species that persist today have known 395 million years of Earth's history. They saw the first trees scale the heavens; they saw the dinosaurs come and go; they saw the first flowers bloom and then swiftly overtake the Earth. There is a sterile hybrid horsetail known as *ferrissii* that cannot reproduce but can only spread like a willow via parts breaking off and establishing elsewhere. Although ancient and impotent, *ferrissii* can be found growing from California to Georgia. Did it cross the country like a newly minted Ph.D. moving to a sprawling technical university and find magnolia trees and sweet tea and black humid nights loaded with fireflies and uncertainty? No. *Equisetum ferrissii* crossed the country like the living thing that it is, and found itself elsewhere, and then did the best that it could.

Part Two

WOOD AND KNOTS

1

THE AMERICAN SOUTH is a plant's idea of Eden. Summers are hot, but who cares, because the rain is generous and the sunshine predictable. Winters are more cool than cold, and freezing is rare. The heavy humidity that chokes us is like nectar to a plant; it allows it to relax and open its pores, and to drink in the atmosphere, confident that evaporation will not interfere. Plants grow all over the South like nowhere else—poplar, magnolia, oak, hickory, walnut, chestnut, beech, hemlock, maple, sycamore, sweet gum, dogwood, sassafras, elm, linden, and tupelo stand over a blanket of trillium, mayapple, laurel, wild grapes, and woefully plentiful poison ivy. In this deciduous world the mild winters are a leafless, lazy time that serves to heighten the drama of the spring explosion of growth. In February, the South begins to burst forth in a profusion of leaves, every single one of which will grow bigger, greener, and thicker through the long, busy summer. In the fall, copious fruit ripens and seeds are scattered, until finally all the leaves are shed in preparation for winter.

If you rake fallen leaves into a pile and then examine them, you will see that each one shows a consummately clean break at the same place near the base of the stem. The fall of leaves is highly choreographed: First the green pigments are pulled back behind the narrow row of cells marking the border between stem and branch. Then, on the mysteriously appointed day, this row of cells is dehydrated and becomes weak and brittle. The weight of the leaf is now sufficient to bend and snap it from the branch. It takes a tree only a week to discard its entire year's work, cast off like a dress barely worn but too unfashionable for further use. Can you imagine throwing away all of your possessions once a year because you are secure in your expectation that you will be able to replace them in a matter of weeks? These brave trees lay all of their earthly treasures on the soil, where moth and rust doth immediately corrupt. They know better than all the saints and martyrs put together exactly how to store next year's treasure in Heaven, where the heart shall be also.

Plants are not the only exploding growth in the American South. Between the years 1990 and 2000, the amount of total income tax collected annually by the state of Georgia more than doubled as Coca-Cola, AT&T, Delta Air Lines, CNN, UPS, and thousands of other recognizable companies relocated themselves to the Atlanta area. Some of this new revenue was channeled into the universities in order to meet the educational needs of a larger and more corporate population. Academic buildings popped up like mushrooms, the number of faculty skyrocketed, and student enrollment continued to climb. In Atlanta during the 1990s, every kind of growth seemed possible.

2

BILL AND I SPENT night after night of those first years designing and redesigning the first Jahren Lab, the same way that a little girl never tires of dressing and redressing her favorite doll. First we put up drywall and split the space into two rooms, each less than three hundred square feet in area. Then we crammed them full-to-bursting with instrumentation—a mass spectrometer, an elemental analyzer, four vacuum lines. We refurbished the ventilation hood to make it capable of tolerating even that most dangerous acid: hydrofluoric. Bill custom-carpentered space-saving compartments under each counter and inside each cabinet that allowed us to fit all of the things we needed and a lot more of what we didn't.

We instinctively hoarded against hard times, which Bill was positive were already on the way. We went to the Salvation Army and got old camping equipment for the lab and amateur oil paintings for my office. We visited the state surplus warehouse, where anyone with a Georgia state employee ID card could help themselves to the mountain of outdated equipment that had been discarded by local government agencies. From there we came home with four 35mm film cameras, an inky mimeograph machine, and two police batons. If we were going to be scientists for fifty more years, we asked ourselves, who could know what might come in handy over such a time span?

One night in early December of that first year, 1997, stands out in my memory, even though it was nearly identical to many other nights that had come before and would come after.

"Season's greetings!" I called out while entering the lab. "What's up?"

Bill's head popped up from underneath the mass spectrometer. "The Elf didn't come by today, if that's what you're asking," he shouted over the noise of the air compressor, which sounded for all the world like an old car with a bad starter. "That goddamn thing is going to make me deaf before my time."

"Huh? What? Huh? Speak up!" I answered.

"The Elf" was what we called the head graduate student within a huge and über-busy lab located on the far side of the campus. Bill had christened the place "Santa's workshop" because of its bizarre atmosphere: upon entering you were surrounded by buzzing students, each of them too preoccupied to even say hello. We ran many gas samples for them, which were delivered to us daily by the Elf.

"If they expect us to work for free, they could at least stick to a schedule," I grumbled.

Bill shrugged. "It's a busy time of year for the Elf," he said, nodding toward the calendar.

"Maybe his heart's not in his work. I've heard that he actually wants to be a dentist."

I wasn't really too concerned: I had finally handed off the revised version of a manuscript to a coauthor, and I was savoring its lifted weight. "Ready for 'lunch'?" I asked cheerfully.

"Sure, why not?" Bill accepted my invitation and we relocated to the microscope lab. "It'll be my treat," he added.

My seventy-pound Chesapeake Bay retriever, Reba, stretched and got up from her basket in the corner. Pleased to see me, she ambled over with her bone, wagging her tail. "Hey, girl, you hungry?" I petted her, rubbing the sharp occipital bone on the top of her head that we referred to as "the Fin of the Beast."

While moving from California to Georgia, I'd gotten lost on the outskirts of Barstow while trying to get myself off of Interstate 15 and onto Interstate 40. Somewhere near Daggett Road, which runs north-south on the eastern side of Barstow, I'd stopped to ask for directions at a parked RV displaying the sign PUPPIES FOR SAIL. When I crouched down and asked the little pack of brown burr-heads which one of them wanted to come to Atlanta with me, a gangly brindled pup had stumbled over with a serious look in her eyes and then tried to climb into my lap. Fifty dollars later (I knew it was meant to be because they took my check), she was my dog.

Like me, Reba spent the best part of her puppyhood in a laboratory, sleeping under the benches and begging Bill to share his tuna-fish-on-saltines dinners. The arrival of each new student spurred the same serious debate between me and Bill as to whether this new person could be anywhere near as smart as Reba. Reba always refused to weigh in, and we weren't sure if she was appalled at our unprofessionalism or convinced that the comparison was facile, or perhaps both.

I took out a small portable TV from one of the cupboards and moved three microscopes to the side, clearing a place for it. In a few minutes it would be 11:00 p.m. and *Jerry Springer* would be on. I put some popcorn in the microwave and cracked open two Diet Cokes. Bill entered carrying nine frozen McDonald's cheeseburgers: three for me, three for him, and three for Reba. He had purchased forty or so of them when the campus commons ran a twenty-five-cent special, and we had happily discovered that their physical properties were not meaningfully changed when reheated from a frozen state.

Bill and I had both left California fairly deep in debt, due to a series of dissimilar but equally foolish purchases from years ago, and had vowed to pay it all off as quickly as possible once we got "real jobs." We soon found ourselves conducting a long-term experiment designed to measure how little we could spend each week and still get by, and frozen food had become a major component of our dietary intake.

We ate in front of the TV, watching a man clad only in a diaper spiritedly invoke First Amendment protection for his "adult baby" lifestyle, waving his bottle for emphasis.

"Man, I would do *anything* to get on Jerry's show," I said with wistful longing.

"So you've mentioned," acknowledged Bill with his mouth full as we watched a montage of the man being changed and powdered by his sweetheart-slash-caregiver.

When lunch was over we cleaned up. "Hey, I've got a wacky idea," I volunteered. "Let's spend tonight running our own samples for a change."

Bill was game. "That's so crazy it just might work," he said, "but first there must be the Airing of the Beast." We went outside and all three of us gazed up at the stars while Bill smoked a cigarette. "This pack cost me more than two dollars," he complained. "I'm going to need a raise."

The entire geographical footprint of a college campus is illuminated all night every night, greatly enhancing its desolation during the weekend. During the school week, the university belongs to no one. It buzzes and throbs with people coming and going. But the whole place is different on a Friday at midnight, when the university belongs to *you*. Smug in the presumption that you are the only working person within a fifty-mile radius, you accomplish just enough to feel justified in being naughty. In the rhythm of these Friday nights beats the honest, humble heart of science, and it also explains how discovery and mischief are two sides of the very same coin.

"A tarnished penny half hidden by the dust of the gutter," I reflected while we were cleaning out the lint traps of the air compressor.

"A penny that will buy you a nonfat soy latte," added Bill, "after somebody loans you three fucking dollars and eighty-four cents."

We had spent the week doing organic carbon extractions, which is a lot more fun than it sounds. For about two hundred million years, dinosaurs roamed our planet in great groups, and a very tiny minority of them were preserved in the mud and silt of their time, including some locked away in Montana until a couple hundred years ago, when landowners stumbled upon them. Dinosaur bones have been carefully excavated, described in painful detail, prepared with special glues, shown to the public, and studied for posterity. Other, less charismatic fossils are of lesser value but potentially greater import, I would argue.

Each streak of brown within each rock that houses a fossil might be the smeared residue of a plant that lived at the time and provided the food and oxygen that supported so many huge reptiles. In these smears there is no anatomy, no morphology, and nothing to take a picture of or to display. However, we might harvest some chemical information from the smear, if we could somehow isolate it and hold it up to the light.

Living plants are distinct from the rocks that surround them in that they are rich in carbon. My colleagues and I decided that if we could capture and separate all the carbon within the dark smears inside the rocks that also held dinosaur fossils, we would have then laid claim to a new sort of plant fossil. The chemistry of this carbon might tell us something about the plant, even though we would never know the shape of the leaves that had made the smear.

In order to liberate the organic carbon—and only the carbon—from a dead rock, we trap the gas that is released while the sample burns. When we do chemistry with liquids, we use beakers to hold one liquid and pour another, to mix the two and to keep others separate. To do chemistry with gases, we use a glass apparatus called a vacuum line, similar to the apparatus that I had been using when I caused the explosion years earlier.

Working a vacuum line is rather like playing a church organ: both have lots of levers to pull and knobs to twist, and it all has to be done in the right order and with the right timing. Both hands are moving at the same time, often performing dissimilar tasks, as the trap and the exhaust are operated independently. After a day of use, both the vacuum line and the organ must be lovingly shut down and delicately maintained; they can both be regarded as pieces of art in their own right. The biggest difference between the two, however, is that a church organ will not explode in your face if you make a mistake while using it.

"Arrgh, I *hate* that thing!" Bill had covered his ears after the loudest air compressor in the world started up with a phlegmatic mechanical cough.

"I know it's awful," I acknowledged, "but a new one is twelve hundred dollars."

"Isn't there someone somewhere who owes us?" he ventured. "Maybe you should write a Christmas letter to Santa."

"*You* are a goddamn *genius,*" I said, and I meant it.

Bill was referring to our ever-increasing exploitation at the mittened hands of "Professor Santa" (the Elf's boss), in which I was supremely complicit, having initiated the whole thing as a barefaced effort to ingratiate myself with an influential person. After reading some of this distinguished professor's publications about oxygen chemistry, I had offered to run some trial analyses of oxygen isotopes for free, and the project had snowballed from there (the puns practically wrote themselves that winter) after he deemed the data to be "very interesting" and then redirected his whole workshop toward making additional samples. We naively consented to run them, vastly underestimating the number of oxygen reactions that could be carried out using a wooden mallet while standing at a conveyor belt and singing.

Earlier that winter, I had gone to a lot of trouble writing multiple private e-mails to the Elf, insisting that he implement a workshop-wide protocol requiring all samples to be labeled in either green or red ink, and bundled in units of ten using silver tape, prior to delivery. My efforts paid off with interest once the sample tubes accumulated to the point where Bill had gotten the joke.

When we inspected our sample logs, we estimated that user "Rudolph" had logged about three hundred free analyses, knowing that they would cost thirty dollars a pop if ordered commercially. We agreed that I would write a letter asking Dear Santa to bring us a shiny, new, silent air compressor. We envisioned ourselves coming downstairs on Christmas morning to find one adorned with a huge red bow, sitting just beneath the biological materials incinerator.

"Start out by explaining how we've been very, very good all year," directed Bill.

"You get the thesaurus and I'll get the department letterhead. It has to be perfect." I was determined to wring as much fun out of the exercise as possible.

"I wonder if they stock crayons in the front office," mused Bill.

While I was digging in my purse for the keys to the office-supplies cabinet, I found a nearly full package of Razzles in one of the pockets, and I stopped in my tracks. "You'll never believe this, but the greatest thing in the history of the world has happened," I told him. We put aside what we were doing, sat on the floor, and divided up the candies, fighting over the precious orange ones and automatically segregating the blue-raspberry ones for Reba because they were her favorite.

The fifty-six hours of weekend that rolled out before us seemed endless. At sunrise we planned to declare ourselves the rightful inheritors of everything in the departmental refrigerator, but beyond that we had nothing scheduled. Maybe we would pick the lock to the machine shop and gawk at the huge saws, drills, and welding tools, treating it as our own personal museum. Maybe we would stage a private showing of *The Seventh Seal* using the projection system in the main auditorium. And maybe there was someone somewhere in the world who was happier than I was during that year, but on nights like that I certainly couldn't imagine it.

3

PLANTS HAVE FAR MORE ENEMIES than can be counted. A green leaf is regarded by almost every living thing on Earth as food. Whole trees can be eaten while they are only seeds, while they are only seedlings. A plant cannot run away from the endless legions of attackers that comprise an unremitting menace. Within the slime of the forest floor thrive opportunists that regard all plants, dead or alive, as nourishment. The fungi are perhaps the worst of these villains. White-rot and black-rot fungi are everywhere, so named because they have chemicals that can do what nothing else can: they can rot the hardest heart of a tree. Four hundred million years of wood, save a few fossil slivers, has decomposed back to the sky from whence it came. All this destruction can be attributed to a single group of fungi that makes its macabre living by rotting the ligneous limbs and stumps of a forest. Yet within this very same group are the best—and really only—friends that trees have ever had.

You may think a mushroom is a fungus. This is exactly like believing that a penis is a man. Every toadstool, from the deliciously edible to the deathly poisonous, is merely a sex organ that is attached to something more whole, complex, and hidden. Underneath every mushroom is a web of stringy hyphae that may extend for kilometers, wrapping around countless clumps of soil and holding the landscape together. The ephemeral mushroom appears briefly above the surface while the webbing that anchors it lives for years within a darker and richer world. A very small minority of these fungi—just five thousand species—have strategically entered into a deep and enduring truce with plants. They cast their stringy webbing around and through the roots of trees, sharing the burden of drawing water into the trunk. They also mine the soil for rare metals, such as manganese, copper, and phosphorous, and then present them to the tree as precious gifts of the magi.

The edge of a forest is a hostile no-man's-land and trees do not grow outside this boundary for a reason. Centimeters outside a forest's border we find too little water, too little sun, too much wind or cold for just one more tree. And yet, though rarely, forests do expand and grow in area. Once within hundreds of years a seedling will conquer this harsh space and endure the requisite years of want. Such seedlings are invariably heavily armored with a symbiotic below-ground fungus. So much is stacked against this little tree, although it does have twice the usual amount of root function, thanks to the fungus.

There is a price: during these first years most of the sugar that the little plant makes in its leaves will go directly into the fungus suckling at its roots. The webbing that surrounds these struggling roots does not penetrate them, however, and the plant and fungus remain physically separate but enjoined by their life's work. They anchor each other. They will work together until the tree is tall enough to fight for light at the top of the canopy.

Why are they together, the tree and the fungus? We don't know. The fungus could certainly live very well alone almost anywhere, but it chooses to entwine itself with the tree over an easier and more independent life. It has adapted to seek the rush of pure sweetness that comes directly from a plant root, such a strange and concentrated compound, unlike anything to be found elsewhere in the forest. And perhaps the fungus can somehow sense that when it is part of a symbiosis, it is also not alone.

4

SOIL IS A FUNNY THING, in that it isn't really anything in and of itself but is instead the product of two different worlds coming together. Soil is the naturally produced graffiti that results from tensions between the biological and geological realms.

Back in California, Bill and I decided that we would teach our soil class differently from the way it was taught to us. Instead of just filling out forms and cataloging data, we would teach where soil comes from and how it forms. We'd make students really look at it, touch it, draw it, and come up with their own improvised labels for what they saw. We developed a teaching protocol that goes like this: we simply pick a place and we dig—we dig the earth open until we can see it whole and naked from top to bottom. We expose what was unexposed, and force its secrets out into the open.

All around us we can point concretely to what's alive—a green leaf, a moving worm, a sucking root. At depth lies cold, hard rock, old as the hills to our right and left, and equally devoid of breath and movement—not alive. Everything that is physically located between the two extremes—alive and not alive—we call "soil." At the top of the soil the influence of the living is most obvious, dark brown rubbed from the residue of dead plants, wilted and rotten, mixed into a slime that seeps and stains everything around it. The bottom of the soil is dominated by the rock's legacy; the waters of the ages have dissolved the rock little by little, churned it into a paste, and dried-wetted-dried it over endless cycles to yield a slag that is distinct from the undamaged rocks that lie below. In the middle, these two substances interact, sometimes blooming into the garish streaks of color that so struck us when we drove through southern Georgia.

Bill was born to tirelessly evangelize soil using his God-given talent for marking the subtleties of chemistry, the shading of colors and the twitches in texture that only he can see from within a hole. He can compare the dozens of soils he carries in his memory with the one before him in bafflingly excruciating detail. All reserve is shed when he talks about soil, and I have watched him deliver many dramatic monologues in Irish pubs (perfectly sober) describing how the discovery of new colors in new combinations underground is what he loves most about his work.

In the summer of 1997, we took a group of five students into the field to teach them how to characterize and map soil. The trip was a first-time experience for four of them, and a repeat trip for the undergraduate who volunteered many hours in my lab each week. Bill had warmed up to this guy, and I had also, and so we invited him to come along on each of our research and teaching excursions.

The best way to prevent people from complaining about food while camping is to force each one of them to take responsibility for one night of cooking, and our undergraduate mascot had eagerly volunteered. Desperate to impress us, he had brought cans, boxes, spices, and a bag of potatoes that he peeled and then boiled until soft, but he was only able to start cooking at about 11:00 p.m., after we finally arrived at the campground.

Boiling water on a campfire is tortuously slow, and so I was dismayed when, after removing the cooked potatoes, he set another huge pot of cold water over the heat to boil. Instead of just doling out the potatoes with a fork, which would have been haute cuisine by our standards, he instead began to mash them, adding the fine-milled flour that had appeared from his backpack. I noted with alarm that the cooking process was starting all over again and I asked him what he was doing. "I am making Hungarian potato dumplings," he explained. "My grandmother used to make them. Trust me, you'll love them."

We ate at about three in the morning. "Hey, you should start going by 'Dumpling'!" I exclaimed as we finally sat down to eat, and the student's face lit up; he was delighted with the professional intimacy that a private joke implied.

"I will *not* refer to him as 'Dumpling,'" Bill said with a masculine scowl as he bent over his soup. He was tired and hungry and not in the mood.

The warm night air was perfectly still and we could hear a chorus of frogs that were croaking somewhere in the darkness. We all ate in silence, stuffing ourselves with the delicious dumplings that had been prepared in ridiculous abundance. While we were cleaning up, Bill was the first to comment. "Good dinner, Dumpling," he said solemnly while gathering up the empty bowls. Whatever the student's real name was, I have since forgotten it, for it was never used by any of us again. I have also never eaten anything that tasted as good as those dumplings, in all my years since then.

We were digging in Atkinson County, which, although it does not seem notable for anything, we referred to as "Nirvana" because of its exquisite soils, unequaled within the forty-nine other states and five continents we have traveled. We found it the same way that we found so many of our teaching sites: from the car window. If you drive across the state of Georgia, from the Piedmont plain near Atlanta southeast toward the Atlantic Ocean, you will find yourself driving along a river of red dust that was rubbed from the residue of what might have been mountains in some ancient geological dream.

While driving on Highway 82 toward the Okefenokee Swamp earlier that year, we had seen what appeared to be buckets of rich apricot paint thrown across a ditch of creamy sand. In those days Bill required a cigarette frequently, and so we were in the habit of stopping often to inspect the landscape. When we had pulled over near Willacoochee, the "paint" turned out to be the rusted-iron band within a rare oxisol soil-type, and we immediately decided to include the stop in our soils course.

When we arrive at a soil site with students, we first unload the shovels, the picks, a tarp, the sieves, the chemicals, and a big blackboard with colored chalk. We dig a hole, down deeper and deeper, until we hit hard rock, careful to stand only on one side so that everyone is looking at the same thing. Once we have dug deeply enough, we dig away from the profile, opening a "pit" that is large enough for three people to stand in, and from which we can make an evaluation as to the lateral continuity of the soil's properties. This digging can take hours, and if the clay is thick or the soil is waterlogged, it is physically exhausting.

Bill and I dig together as sort of a waltz, with one of us "throwing" and the other "catching"—one person chips the ground with a pick, while another positions one shovel below to catch the debris. When that shovel is full, it is swapped for another, and the original one is emptied to the side. Unlike holes that are dug for construction purposes, the fill must be carefully piled to the side in order to keep the bottom clear and provide a view all the way up to the top of the hole. Even as we avoid compressing the soil profile, there are always stray students who we notice standing on the top and looking down on us, and we shoo them away like the chipmunks at the campground. When we ask for digging volunteers, we are occasionally taken up on the offer by what inevitably turns out to be the farm kid of the group. But most students don't really want to dig. In the old days they used to stand idly by, watching us dig for hours, which peeved us. Now they turn sideways, surreptitiously searching for cell phone signals.

Once we can see new earth from top to bottom, we take "pins" (old railroad spikes that we painted bright orange) and insert them at the boundaries between what we think we see as layers. Bill and I argue about the direction of the sun and over whether each detail is real or just a shadow, and we work to convince each other of our opinions like lawyers in a contentious trial with no judge and a bored jury.

Sometimes soil boundaries are distinct, as within a chocolate-vanilla layer cake, and sometimes they are as gradual as the edge-to-middle change of red within one square of a Mondrian painting. Although they form a foundation for all of the data that will follow, the number and placing of these soil "horizons" are the most subjective part of the exercise, and each scientist displays a slightly different style. Some, like me, feel we're creating modern art out of the landscape, preferring the result to be huge and whole, with as few rules guiding the eye as possible. We are known as "lumpers" because we tend to lump the details together as we work.

Others, including Bill, are more like the Impressionists, convinced that each brushstroke must be executed with individuality in order to achieve a coherent whole. They are known as "splitters" because they split the subtle details into separate categories as they work. The only way to do good soil science is to put a splitter and a lumper together in the soil pit and let them fight it out until they achieve something that they both know must be correct because neither of them feels satisfied. Left to her own devices, the lumper will dig for three hours, mark the horizons in ten minutes, and then go on her merry way. Left to his own devices, the splitter will dig a hole and crawl inside, never to be seen again. Thus splitters and lumpers are both productive only when forced into bickering collaboration, and though together they produce great maps, they rarely return from field trips still on speaking terms.

Once the soil horizon demarcations have been successfully negotiated, a sample is removed from each layer, relocated to the tarp, and subjected to a battery of chemical tests to determine acidity, salt content, nutrient levels, and a growing list of field-ready chemical attributes. At the end of the day, all the information is transferred to the blackboard, graphed and drawn, and a long discussion ensues about what the visual properties and the chemical properties, taken together, imply about the fertility of the soil—"fertility" being one of the most grandiose and imprecise terms that science has ever produced.

The ideal educational field trip lasts about a week, with one new soil written up each day, and about a hundred miles of driving afterward, toward another site. Five days and five hundred miles provide enough time and space to give students an idea as to how much soils vary across the landscape, and also to expose them to the thoughtful, ramblin' mind-set necessary for soils work. By the end of the trip they are either in love with the work or utterly turned off, and therefore have probably decided upon a major as well.

By dragging students through dirt for five days I can do something far more important and significant than I can do for them during an entire semester behind a desk, and so Bill and I have clocked tens of thousands of miles on these trips.

Bill is the most patient, caring, and respectful teacher that I have ever seen in action. He will sit with a student for as long as it takes, sometimes hours if needed, in order to help him or her learn just one task. He does the very hardest work of teaching, not just relating facts from a book but standing over a machine and showing how to work it with your hands, how it might break and how to fix it when it does. Students call him at two in the morning when they can't work something, and he wearily comes to the lab and helps them—if he's not already there, of course. He continues tirelessly to coax the slower students toward success, long after I've become frustrated and written them off as not trying hard enough.

Of course, twentysomethings being what they are, most students take Bill completely for granted, save a very few who understand that by the end, their thesis is often as much his as it is theirs. Nevertheless, the most efficient way to get yourself fired from my lab and out on the street is to openly disrespect Bill. You can call me any name you like, but he is your superior and you will remember it and act accordingly. For his part, Bill complains about each student with uniformly wicked contempt and then spends yet another day rescuing them from themselves.

At about five o'clock on that day in southern Georgia—technically the same day we ate the dumplings—we filled in the hole that we had dug and packed up our gear. We stopped in Waycross to replenish the gas tank and our candy supply. While we were debating the advantages of the Hershey bar versus Starbursts, Dumpling approached us and said, "I don't want to go see Stuckie. I'm tired of him. And I think he freaks Reba out."

During each field trip we set aside time for one "enrichment" activity, and Dumpling preferred not to revisit the one we'd made a habit of enjoying during our previous trips to that site. "Stuckie" is a fossilized dog that is on display within a museum called Southern Forest World and is even more unique than it sounds. According to the paleontological expertise that was brought to bear upon the specimen, it is the remains of a dog that ran up a hollow tree "probably chasing an animal" and got stuck and died. The tree petrified while the dog mummified within it, thus preserving for eternity the real-life tableau of a *Tom and Jerry* cartoon.

Stuckie fascinated me and I loved to imagine him as Creon breaking into Antigone's tomb, his face contorted into a grimace of need and regret. When I recalled, however, that Reba always refused to go anywhere near the macabre thing, I realized that from her perspective, Stuckie was a sort of canine poor Yorick whose smell probably inspired unpleasant ruminations about a dog's place in the universe. I made a mental note to apologize later while I watched her mill about near the Dumpster in one of my bright-orange Orioles T-shirts, worn to increase visibility when ambling near the highway.

"I don't know." I thought, hesitating, and then said, "Bill really looks forward to Stuckie."

Bill was ambivalent. "My enjoyment of Stuckie is compromised by your babbling about Greek crap," he noted, "which starts earlier and earlier during each trip, by the way."

"Okay, any ideas about where we should go instead?" I asked Dumpling, and Bill shot me a furious look, enraged that I would do something so foolhardy as let a student chart our course. Tradition dictated that we had to go do some goofy tourist thing before returning home.

"How about that place we always see on the billboard? 'Monkey Jungle'? It looks cool," Dumpling offered.

I threw my backpack in the van and whistled for Reba. "Monkey Jungle it is. All aboard!" I called out to the group.

"Why not? It's only eight hours away," growled Bill while staring daggers at me. I smiled sweetly in return, and once he realized that I was serious, we both got in the car.

Bill does all the driving when we go on the road; he is an excellent driver who merges onto the highway, gets behind the biggest truck he can find, and then follows it at a safe distance for as many miles as possible. I am never allowed to drive because I don't have the patience required by big landscapes; my mind wanders as I drive, and the asphalt road starts to seem more flexible than it really is. My job instead is to talk for hours and dream up scenarios outrageous enough to make Bill laugh, which becomes more challenging as the miles drag on.

I used to think that Bill habitually drove fifty miles per hour because of the responsibility he felt toward our student cargo. But after learning the life history of every motorized vehicle he had ever owned, I later realized he couldn't actually know that they were capable of mile-per-minute travel. Regardless, my attitude had become that I could go anywhere in the world, provided that I was willing to ride shotgun long enough. Once we had agreed to skip Stuckie, there was nothing for it but to get on the highway and drive south.

Ten or so exits north of the Florida border we saw a huge black billboard displaying only two words written in neon pink: BUTT NAKED. It bothered me that I couldn't figure it out. "What does that mean?" I mused aloud in my ignorance. "Is it a bar? Or a strip club? Or a video shop or something?"

"I think it's pretty clear what it means," said Bill. "It means that if you get off the highway, there's something butt naked at or near the exit."

"But I mean, is it a woman or a man or a mole rat or what? Is it even connected to something?" I mused. "Or does it imply that you have the opportunity to get yourself butt naked?"

"It's probably some kind of Gomer-code for something really sick," volunteered a student who was notorious for his derision of all things south of the Mason-Dixon Line.

"Listen," explained Bill, "if you're the kind of guy who's going to pull off of the highway after seeing a sign like that, you're probably also the kind of guy who doesn't *care* what's butt naked on the other end. As soon as you see the words 'butt' and 'naked' you hit the brakes and just go with it."

One of the more politically conscious graduate students tried to stir the pot by asking, "Why are you assuming that it would be a *guy* who is going to a place like that?" Bill shook his head and continued to stare at the road, unwilling to dignify his question with a response.

Fortunately, it wasn't long before a better billboard caught our attention. "Explore Monkey Jungle!" it commanded us. "Where humans are caged and monkeys run wild!" We all shrieked in jubilation.

"We must be getting close," suggested one of the students hopefully.

Bill shrugged. "Well, we are in Florida." We had just passed a sign marking the border and welcoming us to the Sunshine State. The attraction we were headed toward was located near Miami, still about seven hours' drive south of where we were.

Monkey Jungle didn't appear quite so inviting when we pulled into its parking lot that night at 1:00 a.m., given that its lights were off and a heavy link chain bound the handles of the front door. Bill jumped out of the van as soon as he parked, to inspect the sign on the door and also to inhale dried *Nicotiana tabacum* leaves, as he had taken to describing it. The students spilled out of the van like an undone bag of marbles, a few rolling off and becoming unreachable as the majority congregated in place. Bill returned to the group and suggested that we set up our tents on the grassy patch in front of the entrance and sleep until the place opened at 9:30 a.m.

He took a drag of his cigarette. "I figure that somewhere during the course of opening shop they'll bother to wake us up," he said.

Dumpling chimed in. "That way we'll be first in line!"

"I don't know if that's such a good idea," I said. "Don't monkeys crow at dawn like roosters or something?"

"You tell us," said Bill as he rubbed out his cigarette. "You're the one who's been sleeping with a monkey." He was referring to my latest on-again, off-again boyfriend, who was indeed no Rhodes Scholar. I stood there with a smirk on my face while Bill unloaded the coolers and then went to work setting up my tent before unpacking his own, a signal that he had meant no offense. In order to signal back that no offense had been taken, I started digging through the cooler and tried to come up with an idea for dinner.

"Well, it looks like dinner-on-a-stick," I announced, having found extremely little with which to cook us anything.

"That's awesome," Bill said supportively, having finished with the tents in record time. "It's my favorite," he added without sarcasm, and then pulled out an armload of wood and got to work building a fire. It was our custom to visit the campus woodshop before each field trip and to load up the van with the scrap pieces of wood that had been otherwise destined for the pulp bin. Afterward we'd do the same with cardboard taken from the campus recycling center. On the way out of town we'd buy one Duraflame log for each day of the trip and a bunch of random food, and consider ourselves prepared for camping. We'd use these materials each night to build what I called an "Andy Warhol fire," within which we'd use the ever-lit log to ignite a continuous stream of recyclable materials, and the emergent blaze always had a satisfyingly garish result. You could cook over such a fire provided that your sleeves weren't flammable and that you didn't mind if the middle of whatever you were eating was cold and raw.

Dinner-on-a-stick meant that each person found a stick and put whatever they wanted on it and then stuck it in the fire and ate it, and that was dinner. The only rule was that if you stumbled upon something really good, you afterward had to make enough for the whole group, or at least try to make it again and divvy up the result. Dumpling was on a roll during that trip and actually managed to poach pears using a Coke can that had been torn in half and ingeniously skewered on a stick. We all agreed that his Hershey's chocolate–drizzled creation was the absolute pinnacle of camping cuisine, except for his dumplings, of course, and everyone went off to bed happily.

Soon after falling asleep I was unceremoniously awoken by someone with a deep voice and an extremely bright flashlight. I stuck my head outside the tent. "Can I help you, officer?" I asked.

Puzzled to find a reasonably clean and articulate woman instead of a desperately unwashed and incoherent man, the patrolman inquired what we were doing there. I explained our field trip in detail, accentuating my pedagogical duty to fulfill one talented student's specific desire to visit the renowned Monkey Jungle in person before his brief youth had faded.

As so often occurs when I find myself in such situations, the officer's authoritarian skepticism melted into hospitality while I waxed rhapsodic over the rare and peerless Floridian soils. Within a couple of minutes he was offering everything from professional surveillance while we slept to a police escort when we chose to depart for Atlanta.

I declined his assistance gratefully, assuring him that I would indeed call 9-1-1 on the pay phone down the road if I needed anything, and we parted on excellent terms.

After he drove off, Bill stuck his head out of his tent. "That was masterful," he said. "You amaze me."

I looked up at the stars and took a deep breath of the humid air. "Damn," I said contentedly, "I love the South."

The incomparable welcome peculiar to the southern states continued the next morning, when the admissions desk of Monkey Jungle waved our whole group through upon receipt of the insufficient sum of fifty-seven dollars, which represented every last bit of paper money that Bill and I could produce from our pockets. After we walked out of the foyer and through the doors that led to the Jungle we were immediately overwhelmed by the screaming. It emanated from a diverse population of monkey inmates as a large number of them turned their attention toward us.

"Good God, it's just like walking into the lab," said Bill, his face twisted into what I recognized as his pre-migraine countenance.

The room we were in was actually a very large courtyard within the building complex, which had all the architectural panache of your average DMV. In a great arc over the top of the courtyard, long stretches of chicken wire were seamed together and appeared to have been repeatedly reinforced in certain places. *Homo sapiens* visiting the courtyard could walk through the space within a hallway bounded by steel mesh; hence the billboard slogan.

Monkey Jungle was indeed a doppelgänger for my lab, and the more I thought about it, the clearer the comparison became. Perhaps the ambiance had been amplified by a couple of orders of magnitude, but each of our research activities was represented by its simian equivalent within the enclosure. Three Java macaques that had been straining their brains over some problem that they could neither solve nor abandon propelled themselves toward us, supposing that we somehow represented an answer. A white-handed gibbon was draped limply across our walkway, either asleep or dead or someplace in between. Two small squirrel monkeys seemed to be trapped in their own private Samuel Beckett play, caught in a web made of equal parts dependence and loathing. In ironic proximity, two other squirrel monkeys were getting along very, very well by the looks of it.

A single howler monkey sat high on a branch in the back, wailing out the entire Book of Job in his native tongue while periodically raising his arms in an age-old supplication for an explanation as to why the righteous must suffer. A red-handed tamarin crouched in paranoia, rubbing its hands together and scheming toward some sinister end. Two beautiful Diana monkeys meticulously groomed each other while psychologically adrift upon an ocean of boredom. An exhausted cadre of capuchins paced the perimeter, compulsively checking and rechecking the empty feeding troughs for the raisin that they were certain was right there a minute ago.

"Every monkey is some monkey's monkey," I said out loud.

I then happened to notice Bill across the courtyard standing face-to-face with a spider monkey, separated only by a rusty screen. Both of them sported the same hairdo, a three-inch-long dark-brown shiny mop that stuck up in all directions, having been groomed with little more than a few vigorous scratches during the last two weeks. This same shag covered both of their faces, and their lithe limbs hung with an athletic readiness that was only weakly camouflaged by their affected slouches. The spider monkey's dark, limpid eyes were very wide open and his facial expression suggested that he was in a permanent state of shock.

The fascination between Bill and the monkey was so complete that it was as if the rest of the world didn't exist. As I watched I felt the cramps in my stomach that customarily foreshadowed the laughing that continues long past the point of being pleasant or comfortable.

Bill finally stated, without redirecting his stare, "It's like looking in a fucking mirror." I doubled over into a series of helpless guffaws that eventually progressed into a sort of prayer for relief.

When Bill was good and ready, he and the spider monkey parted ways, and we progressed into the final chamber of the Jungle, where a huge gorilla named King sat in a cement hole not dissimilar to those used to impose solitary confinement upon the prisoners of my own species. King's three-hundred-pound frame slumped against the tile as he used one foot to listlessly rub a crayon back and forth across a piece of paper. The walls of the room from which we viewed him were plastered with his finished "paintings," each of which had been executed by King using a similar technique; taken together, they expressed an impressively consistent artistic view.

"At least he's publishing," I observed.

We read a plaque that described the heavy crosses that lowland gorillas must bear within their native Africa, which ranged from poaching to disease, yet it was difficult to imagine any corner of the Congo more dismal than the abject constriction within which King had been impounded in Florida. We read a second, rather apologetic plaque describing how King's overflow art could be purchased in the gift shop and that some of the proceeds were earmarked toward the remodeling and expansion of his enclosure. If King had had a handgun, I was pretty sure that he would have blown his own head off, but seeing that he was armed only with a crayon, he appeared to be making the best of his situation. While I waited for the students to run out of raisins with which to feed the monkeys, I inwardly vowed to stop complaining about my relatively bountiful lot in life.

"Well, I hope that poor son of a bitch gets tenure," sighed Bill from his side of the room.

"Oh, I wouldn't worry about it," I assured him. "It looks like his institution views him as permanent, and he is bringing in money."

Bill looked at me. "I wasn't talking about the gorilla."

While trickling through the gift shop we deposited our last coins in the Plexiglas donation box, but we refrained from using our credit cards to buy one of King's paintings. "I may not know art, but I know what I like," explained Bill as he walked away from the display indifferently.

In the parking lot I instructed the students to use the bathroom now, as we had a long drive ahead of us, and in my head I fantasized about the day after my promotion, when I would commission a T-shirt for myself that read I AM NOT YOUR MOM and begin wearing it to work.

Once we were all loaded into the van and the doors were slammed shut, I took off my hiking boots and cracked open a Diet Coke for Bill. "We went to Monkey Jungle to learn about monkeys, and along the way we learned a little bit about ourselves," I quipped in my most saccharine teacher's voice.

"I fucking *met* myself," mumbled Bill as he craned his neck to look backward while reversing the van out of the parking lot.

Once we had merged onto I-95, I put my feet up on the dash and settled into my familiar role of leading the group in killing time. I was about to incite a semantic argument over whether Monkey Jungle was a jungle *of* monkeys or a jungle *for* monkeys, but decided against it after looking into the rearview mirror and observing that Dumpling was already sleeping like a baby.

5

THE LIFE OF A DECIDUOUS TREE is ruled by its annual budget. Every year, during the short months from March to July, it must grow an entire new canopy of leaves. If it fails to meet its quota this year, some competitor will grow into a corner of its previous space and thus initiate the long, slow process by which the tree will eventually lose its foothold and die. If a tree expects to be alive ten years hence, it has no alternative but to succeed this year, and every year after.

Let's consider a modest, unremarkable tree—the one living on your street, perhaps. A decorative maple tree, about the height of a streetlight—not a majestic maple reaching its full height in the forest—a demure neighborhood tree that's only one-quarter the height of its regal counterpart. When the sun is directly overhead, the little maple in our example casts a shadow about the size of a parking space. However, if we pluck off all the leaves and lay them flat, side by side, they would cover three parking spaces. By suspending each leaf separately, the tree has stacked its surface area into a sort of ladder for light to fall down. Looking up, you notice that the leaves at the top of any tree are smaller, on average, than the leaves at the bottom. This allows sunlight to be caught near the base whenever the wind blows and parts the upper branches. Look again and you'll notice that leaves low in the canopy are of a darker green; they contain more of the pigment that helps each leaf absorb sunshine, allowing them to harvest the weaker rays that penetrate shade. When building foliage, a tree must budget for each leaf individually and allocate for each position relative to the other leaves. A good business plan will allow our tree to triumph as the largest and longest-living being on your street. But it ain't easy, and it ain't cheap.

The leaves on our little maple, all taken together, weigh thirty-five pounds. Every ounce therein must be pulled from the air or mined from the soil—and quickly—over the course of a few short months. From the atmosphere, a plant gains carbon dioxide, which it will make into sugar and pith. Thirty-five pounds of maple leaves may not taste sweet to you and me, but they actually contain enough sucrose to make three pecan pies, which is the sweetest thing that I can think of right now. The pithy skeleton within the leaves contains enough cellulose to make almost three hundred sheets of paper, which is about the number that I used to print out the manuscript for this book.

Our tree's only source of energy is the sun: after light photons stimulate the pigments within the leaf, buzzing electrons line up into an unfathomably long chain and pass their excitement one to the other, moving biochemical energy across the cell to the exact location where it is needed. The plant pigment chlorophyll is a large molecule, and within the bowl of its spoon-shaped structure sits one single precious magnesium atom. The amount of magnesium needed for enough chlorophyll to fuel thirty-five pounds of leaves is equivalent to the amount of magnesium found in fourteen One A Day vitamins, and it must ultimately dissolve out of bedrock, which is a geologically slow process. Magnesium, phosphorous, iron, and the many other micronutrients that our tree needs can be gained only from the extremely dilute solution that flows in between the tiny mineral grains within the soil. In order to accumulate all of the soil nutrients that thirty-five pounds of leaves require, our tree must first absorb and then evaporate at least eight thousand gallons of water from the soil. That's enough to fill a tanker truck. That's enough to keep twenty-five people alive for a year. That's enough to make you worry about when it is next going to rain.

* * *

The life of an academic scientist is ruled by her three-year budget. Every third year, she must solicit a new contract with the federal government. The grant money guaranteed within the contract provides the cash that pays the salaries of her employees; it also provides money to buy all of the materials and equipment that she will use within her experiments and to pay for any travel necessary to complete the research objectives. Universities generally help a new science professor "start up" with a limited sum of discretionary funds—the academic version of a dowry—that support her while she attempts to secure a first contract. If she fails to land a deal within the first two or three years, she won't be able to do the work that she was trained to do, and thus won't produce the scholarship necessary to earn her tenure. If a new professor expects to have a job ten years hence, she has no alternative but to succeed. This is all greatly complicated by the fact that there aren't nearly enough federal contracts to go around.

The type of science that I do is sometimes known as "curiosity-driven research"—this means that my work will never result in a marketable product, a useful machine, a prescribable pill, a formidable weapon, or any direct material gain—or if it does indirectly lead to one of those things, this would be figured out at some much later date by someone who is not me. As such, my research is a rather low priority for our national budget. There is just one significant source of monetary support for the kind of research that I do: the National Science Foundation, or NSF.

The NSF is a U.S. government agency, and the money that it provides for scientific research comes from tax dollars. In 2013, the budget of the NSF was $7.3 billion. For comparison, the federal budget allocation for the Department of Agriculture—the people responsible for supervising food imports and exports—was about three times that amount. Each year, the U.S. government spends twice as much on its space program as it does on all of its other scientists put together: NASA's 2013 budget was more than $17 billion. And these discrepancies are nothing compared with the disparity between research and military spending. The Department of Homeland Security, created in response to the events of September 11, 2001, commands an annual budget that is fully five times larger than that of the entire NSF, while the Department of Defense's mere "discretionary" budget comes to more than sixty times that sum.

One side effect of curiosity-driven research is the inspiring of young people. Researchers generally love their calling to excess, and delight in nothing better than teaching others to love it also; as with all creatures driven by love, we can't help but breed. You may have heard that America doesn't have enough scientists and is in danger of "falling behind" (whatever that means) because of it. Tell this to an academic scientist and watch her laugh. For the last thirty years, the amount of the U.S. annual budget that goes to non-defense-related research has been frozen. From a purely budgetary perspective, we don't have too few scientists, we've got far too many, and we keep graduating more each year. America may say that it values science, but it sure as hell doesn't want to pay for it. Within environmental science in particular, we see the crippling effects that come from having been resource-hobbled for decades: degrading farmland, species extinction, progressive deforestation . . . The list goes on and on.

Nevertheless, $7.3 billion sounds like a lot of money. Remember that this figure must support all curiosity-driven science—not just biology, but also geology, chemistry, mathematics, physics, psychology, sociology, and the more esoteric forms of engineering and computer science as well. Because my work is about learning why plants have been so successful for so long, my research falls within the NSF's paleobiology program. In 2013, the amount of funding that paleobiology gave out for research was $6 million. This is the entire annual budget for all of the paleontology research that happens in America, and the dinosaur-diggers predictably secure the lion's share.

Nevertheless, $6 million still sounds like a lot of money. Perhaps we could agree that one paleobiologist from each state in the country should get a grant. If we divide $6 million by fifty, we get $120,000 for each contract. And this is close to the reality: the NSF's paleobiology program gives out between thirty and forty contracts each year, with an average value of $165,000 each. Thus, at any given time, there are about one hundred funded paleobiologists in America. This is probably not enough to answer the public's many questions about evolution, even if we limit ourselves to the charismatically extinct, such as the dinosaur and the woolly mammoth. Note also that there are a *lot* more than one hundred paleobiology professors in America, which means that most of them can't do the research they were trained to do.

Nevertheless, $165,000 sounds like a lot of money, to me at least. But how far does it really go? Fortunately, the university pays my salary for most of the year (it is very uncommon for a professor to be paid when classes are not in session—that is, all summer long), but it is up to me to secure salary for Bill. If I choose to pay him $25,000 per year (he's got twenty years of experience, after all), I need to request an additional $10,000 to pay for his benefits, bringing the total to $35,000 per year.

On top of this, there's the interesting fact that the university effectively taxes the government for the research that its professors do. So, on top of my request for $35,000, I must request another $15,000 that goes straight into the university's coffers, and I never see a dime of it. This is called "overhead" (or sometimes "indirect cost"), and the tax rate that I indicate above is about 42 percent. The rate of taxation is different at each university, and while it can range all the way up to 100 percent at some of the more prestigious schools, I've never seen it dip lower than 30 percent. This tax is ostensibly used to pay the university's air-conditioning bill, fix the drinking fountains, and keep the toilets flushing, though I feel moved to mention that each of these things works only intermittently within the building that houses my laboratory.

Anyway, the total cost of employing Bill for three years under this pitiful scenario is $150,000, which leaves a whopping $15,000 for all the chemicals and equipment necessary to do three years of state-of-the-art high-tech lab work, or to employ student help, or to do any travel, or to attend workshops and conferences. Oh, and remember—there's only $10,000 of *spendable* money because of the university's tax.

Next time you meet a science professor, ask her if she ever worries that her findings might be wrong. If she worries that she chose an impossible problem to study, or that she overlooked some important evidence along the way. If she worries that one of the many roads not taken was perhaps the road to the right answer that she's still looking for. Ask a science professor what she worries about. It won't take long. She'll look you in the eye and say one word: "Money."

6

A VINE MAKES IT UP as it goes along. The copious vine seeds that rain down from the top of the forest sprout easily, but only rarely take root. Green and malleable, they search frantically for something to cling to, some scaffolding that will provide the strength that they so completely lack. Vines resolve to fight their way up to the light by any means necessary. They do not play by the rules of the forest: they place their roots in one optimal spot and grow their leaves elsewhere, a different optimum, usually several trees over. They are the only plant on land that grows farther sideways than it does up. Vines steal. They steal patches of light left unattended and rivulets of rain. Vines do not enter into apologetic symbiosis, but instead grow bigger at every opportunity, a dead scaffold being just as good as a living one.

A vine's only weakness is its weakness. It desperately wants to grow as tall as a tree, but it doesn't have the stiffness necessary to do it politely. A vine finds its way to the sun using not wood, but pure grit and undiluted gall. An ivy plant sports thousands of rubbery green tendrils programmed to wrap around anything and everything, assuming that whatever each tendril touches is strong enough to support it, at least until something stronger comes along. It is a renegade that can improvise like no other: should a tendril touch soil, it transforms itself into a root; should a tendril touch rock, it grows suction cups and cements them firmly. A vine becomes whatever it needs to be and does whatever it must in order to make real its fabulous pretensions.

Vines are not sinister; they are just hopelessly ambitious. They are the hardest-working plants on Earth. A vine can grow an entire foot in length on just one sunny day. Within their stems gush the highest rates of water transfer ever measured in a plant. Don't be fooled by the few red or brown leaves you find on poison ivy in the fall—the plant is not dying; it's just cheating with different pigments. Vines are evergreen, which means that they never take a day off: no long winter vacation like the deciduous trees that they have laboriously scaled. On top of everything, vines do not flower and bear seed until they reach the open sun above the canopy of the forest, and therefore only the very strongest have ever survived.

In an earthly age when people reign supreme, the strongest plants are becoming stronger. Vines cannot take over a healthy forest; they require a disturbance in order to take hold. Some gash has to create open soil, a hollow trunk, a sunny patch that a vine can come into. People can disturb like nothing else: we plow, pave, burn, chop, and dig. The edges and cracks of our cities support only one kind of plant: a weed, something that grows fast and reproduces aggressively.

A plant that lives where it should not is simply a pest, but a plant that thrives where it should not live is a weed. We don't resent the audacity of the weed, as every seed is audacious; we resent its fantastic success. Humans are actively creating a world where only weeds can live and then feigning shock and outrage upon finding so many. This mixed message is irrelevant: there is already a revolution taking place in the plant world as invasives effortlessly supplant natives within every human-modified space. Our impotent condemnation of weeds will not stop this revolution. We aren't getting the revolution we want: we're getting the one that we triggered.

The vast majority of vines found in North America are invasive species whose seeds were accidentally imported from Europe and Eurasia along with tea, cloth, wool, and other basic necessities. Many who immigrated to America during the nineteenth century built a spectacular fortune in a new land. Freed from the torment of insects that had exploited their weaknesses generation after generation for millennia, these vines also flourished in the New World unfettered.

The vine that we know by the name "kudzu" arrived in Philadelphia as a gift from Japan to honor the 1876 centennial. Since that time kudzu has expanded to cover a total land area the size of Connecticut. Thick ribbons of kudzu embroider thousands of miles of highways in the American South. Kudzu thrives within the roadside ditches where we throw our beer cans and cigarette butts: it is the living garbage of the plant world. Kudzu is perpetually where it should not be, blocking our view of prettier pink dogwoods. If we were to wade through the refuse and tease one out, we'd see that a single strand of kudzu can grow to be one hundred feet long, easily twice the height of the forest. Kudzu is resigned to its lot as a parasite; it knows no other way. While the dogwood tree blooms, stationary and secure in its expectation of another glorious summer, the kudzu resolutely continues to grow one inch each hour, searching for its next temporary home.

7

AFTER WE FOLLOWED Dumpling's directive to Monkey Jungle and received the epiphany that we were all just monkeys working in a monkey house, everything started to make sense. When I was separated from the lab, attending some seminar or conference, it was the series of twisted e-mails from Bill that held me fast to what I loved about my job, even while trapped with pasty middle-aged men who regarded me as they would a mangy stray that had slipped in through an open basement window. *There's a place somewhere where I am part of the in-group,* I would remind myself as I stood alone with my buffet dish in some Marriott ballroom, apparently radiating cooties and so excluded from the back-slapping stories of building mass spectrometers during the good old days.

Each time I returned to Georgia Tech from traveling, I tried to throw myself into working even harder. I began to set aside one night a week as an all-nighter (Wednesdays) in order to complete the paperwork that went unattended while I served on committees tasked with documenting the potential obsolescence of chalkboards on campus. I learned that female professors and departmental secretaries are the natural enemies of the academic world, as I was privileged to overhear discussions of my sexual orientation and probable childhood traumas from ten to ten-thirty each morning through the paper-thin walls of the break room located adjacent to my office. By these means I learned that although I was in desperate need of a girdle, I was better off than one of the other female professors, who would never lose all that baby weight by working all of the time.

As hard as I worked, I just couldn't get ahead. Showers became a biweekly ritual. My breakfast and lunch were reduced to a couple of cans of Ensure from the cases that I kept under my desk, and in desperation, I once threw one of Reba's Milk-Bones in my purse so that I could gum it during a seminar, trying to keep peoples' attention off of what I knew would be my growling stomach. The acne that I had never wrestled with as a teenager decided to make up for lost time with a magnificent debut, and I passed the workday biting my nails with ferocity. My brief forays into romance had convinced me that I would be relegated to love's bargain bin; none of the single guys that I met could understand why I worked all of the time, and nobody wanted to listen to me talk about plants for hours, anyway. Everything about my life looked pretty well messed up compared with how adulthood had always been advertised to me.

I was living on the outskirts of town, just where Atlanta ends and southern Georgia begins. I rented a trailer that presided over three less-than-pristine acres of Coweta County and paid extra for the additional privilege of caring for a geriatric mare named Jackie. I figured that this was worth the thirty-five-minute commute: I had always longed for a horse, and the fact that I was officially done with school and employed made it seem possible. Jackie was lovely and a consistently soul-nourishing source of equine comfort to me and made fast friends with Reba. My only complaint was that both my neighbor to the west and my landlord had progressed from being friendly to being creepy as soon as I got my bags unpacked.

It puzzled me that the makeshift garage of the trailer contained stacks and boxes overflowing with homemade VHS tapes. My landlord had given me some lame excuse as to why all these tapes couldn't be kept at his house, and I had shrugged and shut the door, not needing the space anyway. The more I thought about it, however, the more difficult it became to conjure up an innocent reason for him to have so freakishly much video stored well away from his wife and children. He was also always showing up unannounced and regaling me with how fascinated he had become that such a little slip of a thing like me was willing to live alone way out in the woods with no gun at her disposal.

In a similar vein, my neighbor to the west took to passing by of an evening in order to assure me that, although he may not have looked like he was up to it, his EMT training conferred upon him the necessary skills and experience with which to cut off my clothes in less than forty-five seconds should he deem it necessary. I eventually learned that in Georgia, when someone walks up to you wearing overalls with no shirt underneath them, it is unlikely that something good is about to happen.

After one year, the "check engine" light lit up on the first car that I'd ever owned. Because I had no idea what it meant, I took it as a sign and traded the damn thing in for a used Jeep, loaded up my dog, and moved into town. I secured lodgings: a long, narrow basement apartment that Bill promptly christened the "Rat Hole" within Atlanta's Home Park neighborhood. The Rat Hole butted up against the stockyard of a working steel factory, and I learned many interesting things, including the fact that the fabrication of steel involves dropping whole reams of metal sheets from twelve feet at regular intervals throughout the night. I spent countless humid Georgia evenings sitting on the back steps at the entrance of the Rat Hole, watching Bill's cigarette tip glow among all the other blinking fireflies and trying desperately to formulate some kind of Plan B against the background music of industrial drums marching me inexorably toward menopause.

Bill's lot was considerably weirder than mine, though he endured his with rather more nonchalance, and much more resilience, than I did. He had landed in Atlanta, happy to find that the monthly rent for a filthy firetrap in Georgia equaled one-tenth the monthly rent for a filthy firetrap in California, but after going ten rounds with Confederate bedbugs he was eager to declare surrender if not outright defeat. He bought a Volkswagen Vanagon (goose-shit yellow) and I helped him move into it, culminating in the strange experience where one loads up one's belongings and drives off to . . . well, nowhere, seeing as one is already precisely at home.

Before we had gone a block toward nowhere, we heard a hard thump followed by the yowl of a cat and we knew that we were passing the "Felisphere," a fully functioning feline ecosystem that we had named after Columbia University's Biosphere project in Arizona. It was an old house inhabited by hundreds of apparently self-sufficient cats that patrolled the neighborhood, their activities disrupted only superficially by human traffic. I forced Reba to duck down in the backseat, knowing that canine hubris is never more tragic than in the face of superior numbers.

"Those cats never liked me," Bill reflected. "They never wanted me to move in." He stuck his head out the window. "So long, you furry assholes," he shouted. "You won't have my shoes to piss in anymore."

Bill was hard to find when he was living in the van: cell phones weren't common then, and, by definition, he didn't have a fixed address. If he wasn't in the lab, I simply had to go cruising around for him. I'd check the usual hangouts, knowing that if I did see the van, he was unlikely to be very far away.

"Welcome. Can I offer you a hot beverage?" Bill greeted me from a posture of repose when I walked into the coffee shop that he considered his "living room." The place was located next door to a Laundromat ("my basement"), and he could reliably be found there on Sundays. On that particular morning, he was sitting comfortably in a plush armchair reading the *New York Times* in front of a gas-powered fireplace with a double latte in his hand.

"You cut your hair again. I hate it," I observed.

"It'll grow back," Bill assured me as he rubbed his head. "It was just one of those Saturday nights, you know."

There were certain things in life that Bill would go to almost any length to avoid, and one was going to a barbershop. The very idea of the physical intimacy inherent to the hair-cutting process overwhelmed him, and from the time that I had met him in California, he had sported long, glossy black hair that was not unreminiscent of Cher's. He was commonly mistaken for a woman from the back and endured constant suggestive sidelong glances from passing men, which then morphed into embarrassed and resentful surprise once they got full sight of his scruffy beard and masculine jaw. This did nothing to decrease Bill's social paranoia, and not long after moving into the van, Bill bought a cordless electric razor—the kind you find at a real haircutter's. He called me at 3:00 a.m. one night about a month later and excitedly relayed that he had shaved his head.

"It's very liberating; I feel great. Long hair is so fucking foolish, I feel sorry for guys who have it," he said, expressing the total conviction one finds in those who have only recently been converted.

"I can't talk right now," I stammered, and hung up nervously. I didn't like the idea of Bill changing drastically, and this was too much for me to absorb. Would Bill without all that hair still be Bill? I knew it was irrational, but I felt the need to avoid him for several days. I would see him soon and take it all in, just not right away. I kept telling myself that the shock would be too much, and so I kept making excuses and lying low. Bill noticed this, of course, and it confused him.

Eventually he called me from a pay phone in the middle of the night. As soon as I picked up he said, "I still have the hair, you know. Would it make you feel better to see it?"

I thought about it, and decided that it probably would. "It's worth a try," I agreed. "Come pick me up."

Bill arrived in the van, and I got in, avoiding eye contact. "It's at the reservoir," he explained as we turned north onto Howell Mill Road. Finding a place to park the van for the night was an aggravating problem that Bill had to solve on a daily basis. This was greatly complicated by the fact that the van only barely ran, and so parking and breaking down were pretty much the same thing.

Several factors contributed to the complexity of the problem. The van didn't have reverse, so any parking space had to be pull-through. If someone blocked you from the front you were pretty much stuck there for as long as they stayed, and you had to guess where other people were going to park. The van also lacked first gear, so a slight incline was needful, as a rolling start was imperative to getting the van going in the morning. Worst of all, the starter didn't function once the engine was warm, so wherever you turned off the van, you had to stay for at least three hours until the whole thing cooled down enough to start up again. Fueling the van was a dicey arrangement, since the engine couldn't be turned off while filling. Pumping gas is not normally much of a high-adrenaline activity, but watching Bill slop the nozzle over the van's sparking muffler with a cigarette hanging out of his mouth could really get your pulse racing.

It was about four in the morning when we arrived at the reservoir's overlook, which wasn't much to look at, over or otherwise, truth be told. Bill drove onto a small hill and stopped the van (though not the engine) at a slight incline pointing down. "This okay?" he asked, his hands on the keys. He was asking me in code if I thought this was a good place to hang out for three hours before he shut down the engine.

"We went to the reservoir to live deliberately." I signaled my assent by misquoting Thoreau. Bill usually referred to the reservoir as his "weekend getaway," since it was technically water surrounded by trees and was ill policed during the weekends. In the harsh light of day it was an ugly square reservoir surrounded by a twelve-foot fence rusted through in places, plus a few odd scraggly trees that had been overgrown by kudzu.

Bill turned off the van, took his keys out, and pointed with them, straight ahead. "The hair's in there," he told me.

"Where?" I asked, not sure toward what he was pointing.

"In *there,*" he repeated, pointing specifically toward the big sweet gum tree situated about ten feet in front of the van. I got out and walked over to it. I realized he was probably talking about one of the hollowed-out chambers in its trunk.

"Just reach in, it's right there," he encouraged me.

I stood and considered it for a while. "No," I declined, "I don't think I will."

"What is your problem, anyway?" Bill said in exasperation. "You're acting like a guy shaving off his hair and then hoarding it in a dead tree on the wrong side of town isn't a totally normal thing. My God, you are hung up."

"I know, I know," I confessed. "It's not you, it's me." I was quiet for a while, probing my subconscious. "I guess I just don't like the idea of that big a part of you just getting cut off and thrown away," I explained, as well as I could.

"Duh! Duh!! *Duh!!!*" Bill exclaimed. "Neither do I! Of course fucking not." His voice was tight. "That's why I am storing it here. I'm not a barbarian, for Chrissakes." He reached into the hollow and pulled out a huge wad of black hair. He held it up and shook it under the light cast by the buzzing fluorescent bulb that was perched atop a well-graffitied pole.

I stared. "It is magnificent," I had to concede, and rapturously. I was impressed by both the glossiness and the sheer volume of the snarl; from a distance it might have looked as if he were waving goodbye to someone with a dead cat.

We looked each other in the eye and laughed. From then on, when he shaved the hair from his head he always stuffed the product into the same tree and we would visit it occasionally, late at night. It was a comforting ritual, although I was sure that one of us would eventually get bitten on the hand by a raccoon while reaching in.

On the nights when we visited the hair, we'd often sit at the reservoir and brainstorm about a children's book based on Bill's life, which we both agreed comprised the most gleefully inappropriate material for such a thing. This particular installment would be called *The Getting Tree,* and it was about an arboreal parent figure that slowly cannibalized its offspring because of its progressive and oblivious greed. In one of the middle chapters the Boy visits the Tree soon after entering puberty, hoping to find within its arms a retreat from the vicious world of adolescence. "I see you're getting hair on your chest," says the Tree. "Shave it off and give it to me," it demands casually.

Toward the end of the story the Boy has turned into a very old man and has gone completely bald with age and worry. "The raccoons are having baby raccoons again; I need more hair," says the Tree. The Boy shakes his head apologetically. "I am sorry, but I have no more hair to give you, being just a bald old man." "Stick your arm in the hollow, then, and the raccoons will chew it. An old man's arm is good for chewing anyway," suggests the Tree. "Yes," agrees the Boy. "Let us just stand together then, and I will lean upon you and let them chew awhile." As the book closes, its action is resolved into a poignant tableau of sacrifice.

"That's pure Caldecott, right there," I observed one night after a particularly productive editing session.

Less than six months after Bill started living in the van, he knocked on my door at three-thirty in the morning. He came in, and I went to get him some of the coffee that I had just started to brew.

That summer had not been going well. "It's a hard life, living in a van," Bill would say often, with a wistful sigh. The Georgia heat commonly exceeded ninety degrees by eight-thirty in the morning, making sleeping inside a car until any normal hour completely impossible. Bill combatted the heat with ingenuity: he found a parking spot within campus lot P3 where he could position the van sideways beneath the drape of an overgrown willow tree and obtain both cover and shade. He blacked out all the windows, including the windshield, with aluminum foil placed with its reflective side facing out. This kept the van bearably cool until the sun had fully risen.

I would cross paths with Bill as early as 7:30 a.m. and see him staggering groggily around the lab with a beaker of water in each hand. By his telling, he had been "baked out" about an hour earlier and was, as usual, "exceeding parched." The desiccating nights he experienced were exacerbated by his habit of ceasing fluid intake at about 6:00 p.m. each evening; he had no options for urination, and he scoffed openly at the idea of using the bushes. "I have my standards," he declared haughtily.

On the night that he showed up at my house, his sleep had been dramatically interrupted. We had always marveled that no one seemed to notice or care that Bill's creepy van was perpetually parked in P3, but in the end it turned out that someone had and did—namely, the campus police. One night, while Bill was sleeping in a pool of his own sweat, he awoke to an energetic pounding on his windshield. Outside he could hear a squad car siren chirping against staccato CB radio communications. He rolled the van's door open.

He didn't look like a particularly model citizen: he had planned to shave his head the day before but got only halfway done before the batteries in his razor ran out, which gave him a sort of escaped-mental-patient look. The van stank in the way that such close quarters usually do, and across the passenger seats were sprawled the guts of his portable television; he had taken it apart in order to tinker with the wiring. While a flashlight blinded him, he heard a disembodied voice asking, "Sir, can we see some identification?"

After they were satisfied that there was nothing sinister in the van, Bill showed them his driver's license, university ID, passport, and even the ziplock bag of hair recently harvested from the left side of his head. Soon after that, I received a call from the police asking me to verify that Bill was my employee.

"We found him sleeping in a van in a campus parking lot," they explained to me over the phone.

"Yeah, lot P3," I confirmed. "Under the willow."

Once they figured out that Bill represented no threat to anyone, and that he certainly hadn't done anything criminal, the police officers were extremely apologetic for disrupting his evening. They really had no choice but to wake him up; it was their job, and, well, you know how it is. Bill assured them that there were no hard feelings. "You know there's a campus emergency phone right down the hill," one of them reminded him in a fatherly way. "You be sure to use it if you ever need anything." After they drove off, Bill dressed and came over to my place, supposing that I might appreciate an explanation for the phone call.

"I don't know how you can be so calm about this." I was upset. "You are exactly the kind of guy that they could pin something on if they wanted to . . . a weirdo loner who periodically shoves body parts into a tree?"

"Oh, come on . . . I have nothing to hide. I don't do drugs and I don't make trouble. I positively radiate normalcy," he said, and I had to agree that it was true, in its way. Neither of us had ever done any drugs, even during all those years at Berkeley. In fact, we didn't even drink beer on field trips, which was practically unheard-of in the earth sciences. I had knowingly made some photocopies under the previous user's departmental code, but I hadn't done anything worse than that so far that semester.

"Well, you do swear too much," I countered, unwilling to completely concede his point. Bill agreed that this was probably fucking true. "And look at you: you look like the second coming of Eraserhead; you're lucky they didn't haul you in just for that." I was angry and scared.

Then I relented. "Listen, I know this is all my fault. It's because I don't pay you a living wage. But I can't—at least not yet. But soon—soon, I think—we're going to get a really big grant." I searched for something to say that would make my promise sound less empty.

"Anyway, this was the last straw," I told him. "I'm tired of worrying about you every night. You've got to find somewhere to live." I wracked my brain for a solution. "I'll give you the money."

Bill did find somewhere to live. During the next week he moved into the lab. He slept in one of our student offices—the one that no one wanted to use or even wanted to enter. It had no windows and no ventilation and thus had absorbed the body odor of everyone who had ever worked in the building, fermented it within the ceiling tiles, and continuously exuded it as a rare bouquet. He called it "the Hot Box" because it was perpetually five degrees hotter than the rest of the well-heated and poorly cooled old building.

He improvised a bed and dresser behind the cover of an old desk and took to sleeping in a T-shirt and khaki pants (his "pajakis") so that he could rise up immediately if a secretary or janitor entered, claiming that he was just resting his eyes midway through a long lab experiment. This was nearly ideal, except for the fact that the Hot Box was located near the front entrance of the building, and Bill found it especially hard to sleep after 9:00 a.m. once the hordes entered, swinging the doors open and shut. He replaced and greased the relevant hinges, but it didn't help much. After one particularly late night, he put up signs that read DOORS BROKEN, PLEASE USE BACK ENTRANCE, but that lasted only until Facilities was called over and couldn't find a problem.

He packed the biological sample freezers full of frozen dinners and kept his bulk groceries stored within the secretaries' fridge until they complained about the three whole watermelons that had proven irresistibly cheap at Kroger. Taken all together, Bill seemed pretty content except for one thing: a lack of private showering facilities. He rigged up a sort of bidet within the mop sink of the janitor's closet, but he had to leave the door propped open so that he wouldn't get locked in while he was using it. Try as we might, we couldn't come up with a convincing cover story for why he would be in there, soaped up and naked at 3:00 a.m., and I think this fed his natural tendencies toward paranoia.

One morning at about eleven o'clock, the fire alarm went off in the building, and upon leaving my office, I saw Bill shuffling along with the many others involved in the evacuation, barefoot in his pajakis with his hair sticking up in all directions and a toothbrush hanging out of his mouth. Once he got outside, he stumbled over to a windowsill planter of geraniums and spit toothpaste into it.

I walked over and greeted him. "Dude, yuck. You look like Lyle Lovett out on a day pass from somewhere."

Bill began to repeatedly flick his near-empty lighter, trying to get one last flame out of it. "If I had a boat," he mumbled around his cigarette, "I'd go out on the ocean."

Because he literally had nowhere else to go, Bill was working in the lab for about sixteen hours a day. By virtue of availability, he soon became everyone's counselor and confidant. He would help the students fix their bicycles and change the oil in their old cars, go over their 1040EZ forms with them and help them figure out where to show up for jury duty—grumbling about it all the while. When the students told him about their lives, in the charming way that only a nineteen-year-old undergraduate does ("Get this: the closet in my dorm room has a *built-in* ironing board!" "Can you believe it? I'm going to assistant-produce the campus radio station's Sunday-morning 3:45 a.m. post-reggae-punk music hour!" "At Thanksgiving, when my dad said he had never heard of Gertrude Stein, I was like, 'Who *are* these people?' "), he would listen and never judge. He also never reciprocated with any stories about himself, but the students were too absorbed with being young adults to notice.

As a rule, Bill didn't share the students' stories with me, but he did make sure to pass on the best of the best. Karen was an undergraduate lab assistant who wanted research experience on her résumé in order to beef up her application to veterinary school. Ultimately, she wanted to work with endangered animals that had been rescued from captivity and help repatriate them to their native surroundings.

She left us for the summer in order to accept a coveted internship at the Miami zoo, only to find that most of what zookeepers actually do amounts to pretty routine hygiene maintenance, and that the only thing worse than an animal that doesn't appreciate this is one who does.

Placed upon the lowest rung of the ladder, she was sent to work in the primate enclosure. Karen's job was to apply anti-inflammatory cream to monkey genitalia, which were in need of daily soothing due to their constant and indiscriminate use. Once the monkeys had recognized her as their new vehicle of relief, they began mobbing her when she entered the room. Bill and I could hardly absorb this story when she told it to us, it was just too wonderful, but it got even better. It turns out that it is a hard-hearted monkey indeed that remains unmoved during a good slathering of bacitracin, and most monkeys proved considerably more responsive to her reluctant manipulations.

The zoo had fitted Karen with a protective plastic shell meant to discourage her charges from clutching on to her and wildly humping her frame, but it wasn't 100 percent effective. On the upside, her many animal behavior classes had provided her with the intuition necessary to condition these monkeys to the concept of a glory hole; the downside was that seeing them lined up and "standing at attention" through a chain-link fence first thing in the morning was enough to make her rethink a career in veterinary medicine altogether. She returned to our lab after the internship having decided that maybe botany wasn't so boring after all.

Even though we were always on campus, we didn't know everyone. There was a strikingly pale fellow who used to attend the weekly seminar regularly, always sitting alone, far back in the last row to one side. His countenance was waxy white, and his hair was long and white too, though he didn't look to be more than middle-aged. He would slip into the lecture hall at the last moment and be the first to slip out at the end, skipping any and all refreshments and conversation. We never saw him otherwise, and we never heard him speak a word nor saw him interact with anyone. We decided that he lived in the attic of the building and started calling him "Boo Radley." I tried to follow him one day, dodging out early during the questions session so as to be ready, but he somehow lost me during the confusion of the mass exodus.

I used to speculate endlessly about Boo—his probable reactions to each seminar, his expertise, his personal fortune—and then contrive tactics by which we could expose him, violate his privacy, and discover everything that I wanted to know. Bill never showed any interest in my schemes. One night, he sat calmly on the building's front steps as I pressed him on the subject, pointing excitedly to the one light that still glowed out of a third-floor office.

Bill looked up at the light and then out to the stars. He took a deep drag on his cigarette, exhaled, and said, "I don't know, Scout. He is who he is. I think I'd rather not know more than that. It's enough to know that he's up there, and that he'll step in and save us if anything really bad ever happens." Bill crushed his cigarette on the pavement, looked at me, and took off his fleece jacket. He handed it to me so I could put it on before I even realized that I was cold.

8

A CACTUS DOESN'T LIVE in the desert because it likes the desert; it lives there because the desert hasn't killed it yet. Any plant that you find growing in the desert will grow a lot better if you take it out of the desert. The desert is like a lot of lousy neighborhoods: nobody living there can afford to move. Too little water, too much light, temperature too high: the desert has all of these inconveniences ratcheted up to their extremes. Biologists don't much study the desert, since plants represent three things to human society: food, medicine, and wood. You'll never get any of those things from the desert. Thus a desert botanist is a rare scientist indeed and eventually becomes inured to the misery of her subjects. Personally, I don't have the stomach to deal with such suffering day in and day out.

In the desert, life-threatening stresses aren't a crisis; they are a normal feature of the life cycle. Extreme stress is part of the very landscape, not something a plant can avoid or ameliorate. Survival depends on the cactus's ability to tolerate deathly grim dry spells over and over again. If you meet a barrel cactus that's tall enough to touch your knee, it is likely to be more than twenty-five years old. Cactuses grow slowly in the desert—during the years when they do grow, that is.

A barrel cactus has folds like an accordion, and deep within these folds are the pores that let air in and water evaporate out. When it becomes very dry, a cactus sheds its roots to prevent the parched soil from sucking all the water back out of it. A cactus can live for four days with no roots and still continue to grow. If there is still no rain, the cactus begins to contract, sometimes for months, or until all the folds have closed together. Its spines form a dense and dangerous fur protecting what is now a hard, rootless ball of plant. In this posture, the cactus can sit without growing and await rain for years, while continuously punished by the sun. When it finally rains, the cactus will either return to full functioning within twenty-four hours or show itself to be dead.

There are a hundred species or so known as "resurrection plants." These species are unrelated, but within each of them the same process has somehow developed. Resurrection plants have leaves that can be desiccated to papery brown shreds, feign death for years, and then rehydrate back to normal function. It is their unusual biochemistry that allows them to do this, an accidental trait and something that they did not choose. As they wither, their leaves fill with concentrated sucrose, thick sugar left behind during the drying. This syrup stabilizes and preserves the leaves, even when they are drained of their green chlorophyll.

Resurrection plants are usually tiny, no bigger than your fist. They are ugly and small and useless and special. When it rains, their leaves puff up but do not become green for forty-eight hours because it takes time for photosynthesis to start up. During those strange days of its reawakening the plant lives off of pure concentrated sugar, an intense sustained infusion of sweetness, a year's worth of sucrose coursing through its veins in just one day. This little plant has done the impossible: it has transcended the wilted brown of death. The miracle is not sustainable, of course, and within a day or two things will inevitably go back to normal. Such a crazy life takes its toll, and in the long term, even a resurrection plant withers and dies completely. But for a brief, glorious moment it knows something that no other plant has ever known: how to grow without being green.

9

FULL-BLOWN MANIA LETS YOU SEE the other side of death. Its onset is profoundly visceral and unexpected, no matter how many times you've been through it. It is your body that first senses the urgency of a new world about to bloom. Your vertebrae seem to detach from one another and you elongate as if toward the sun's light. You can't hear above the sloshing roar of blood pushed through your head by some impossibly sustained orgasm within your beating heart. For the next twenty-four, forty-eight, seventy-two hours you will have to yell to hear yourself over this whooshing. Nothing, nothing can be loud enough or bright enough or move fast enough. The world appears as if through a fish-eye lens; your view is fuzzy with sparkling edges. You have received a grand, systemic injection of Novocain and your entire body tingles briefly before it becomes flaccidly foreign and unreal. Your raised arms are the fleshy petals of a magnificent lily bursting into flower. It deeply dawns on you that this new world about to bloom is *you*.

Deep night is no longer dark and why did you ever think it was dark at night? The darkness of night is like all of the other impossibly simple things that you used to believe but against which there is coming a revelation of multidimensional glory. Soon you don't register day *versus* night because you don't need sleep. You don't need food or water or a hat against the frigid weather, for that matter, either. You need to run. You need to feel the air on your skin. You need to take off your shirt and run so you can feel the air and you explain this to the person holding you that it's okay it's okay it's okay to do this but he doesn't get it and his face looks worried like someone died and you feel pity for him because he doesn't realize how wonderful and okay and okay and okay everything is.

So you explain how it is and he doesn't get it and you tell more and more in a different way and he isn't really listening and he says don't you have anything for this and why don't you take some of it and you explain that you don't want that stuff you need to feel this and he doesn't get it and he doesn't get it and so you viciously order him to go go go away. And he finally does. But it's okay because you didn't mean it and you'll explain it all later and he'll see because it will all make sense and he'll be glad too once he understands that something wonderful is about to happen and it would be a sin to keep it from coming.

Then comes the best part. It is the final lifting. Not only has the weight of your body dissolved but all the collective trouble of this ancient, weary world as well. The hunger, the cold, the misery, the hopelessness of every human hurt the world over seems manageable and solvable. There is nothing, nothing, nothing that will not be transcended. And you are the exalted, the one person out of billions utterly free from the burden of the existential pain that all must carry. The future will be splendid and full of miracles and you can taste its coming.

You don't fear life and you don't fear death. You don't fear anything. There is no sadness and there is no grief. You feel your subconscious formulating the answers to all the collective miserable searching that man has ever done. You have indisputable proof of God and the creation of the universe. You are the one for whom the world has waited. And you will give it all back; you will pour out all you know and then wallow knee-deep in thick viscous love, love, love.

When I die I will identify Heaven only secondarily by these feelings, and primarily by their failure to end. While I am constrained to this life they will always end, and what comes after, like any resurrection, is not without cost.

While this great cosmic fire hose bathes you in epiphanies, you are overtaken by your urgent need to document them and thus generate an inspired manual for all perfect tomorrows. Unfortunately, this is also when reality closes ranks and conspires to thwart you in earnest. Your hands shake such that you can't hold a pen. You pull out a tape recorder and push "record" and fill cassette after cassette. You talk until you are hoarsely coughing blood, you pace like an animal confined until you faint. Then you get up, change the cassette, and continue on because you are so close to something, some proof or some desperate hope that your own little life was actually meant for something less confusing and more worthwhile.

And then it's too loud and it's too bright and there's too much too close to your head and you scream, scream, scream it away. And then someone is holding you saying oh god how did this happen and what is this hair and sweet jesus one of your teeth on the floor and they dab at the blood and snot. And they feed you a single sleeping pill and you sleep and wake to one more sleeping pill and you sleep and wake to one more sleeping pill as if they were feeding with an eyedropper a broken baby robin that had fallen out of its nest. Hours or is it days later you wake to a gray sadness that mutes you into a silent weeping numbness and you wonder why, why, why you are being punished like this.

Finally fear overcomes sadness and you roll back the stone, crawl out of the tomb to assess the damage, and then do what needs to be done. Fear overcomes shame and you make a doctor's appointment to beg and wheedle for more sleeping pills as your only stockpile against next time, next time, next time.

And by luck, by stupid luck, or time or chance or Providence or Jesus or who cares, your appointment happens to be at the best hospital in the world and a doctor looks at you hard and he says, "You don't have to live this way." And he asks questions until you've told him everything and he's not horrified or disgusted or even surprised; he says people have this and they manage it. He asks you how you feel about medicine and you tell him that you aren't afraid of anything made in a laboratory. He smiles and describes the drugs one by one and you want to get on the floor and kiss his hand like a dog because you are so grateful. This doctor is so smart and so sure and has seen this so many times that you begin to dare to hope that maybe it's not too late to finally grow into what you were supposed to be.

Years later, while preparing to move across the world, you find the stack of cassette tapes at the bottom of your closet. You realize that you won't be bringing them along. One by one you disembowel the tapes, pulling out reels of shiny brown floss. A curly mess is all that remains from the sprawling ecstasy of those anguished high holidays. You sit for an hour and vow to try to love what's left of the poor sick girl who recorded her own cries night after night with only a machine to hear her. You decide that this plastic snarl, though dead, is still precious, that it is the placenta that was attached to you while you writhed in the dark waiting to be born. You stand up, carry it outside, and bury it under the magnolia tree. You come inside and pack all of the things that you will take with you, and try to forgive yourself for what you're leaving behind.

But that particular day of health and healing is still many years distant within my story, so let's go back to 1998 in Atlanta and I'll keep describing how the world spins when mania is as strong and ever-present as gravity.

10

"WHERE THE HELL have you been?" Bill hollered at me when he came around the corner and saw me standing in the lab.

I blinked at him numbly. "I've been in a funk," I tried to say offhandedly even as I choked on my shame. I had been lying in bed crying for thirty-six hours after crashing down from my latest throbbing mania, this one triggered by the corticosteroid injections necessary to quell an acute allergic reaction. We had been studying plants along the Mississippi River, traveling through Arkansas, Mississippi, and Louisiana while trying to sample our way through an unbelievably lush gauntlet of poison ivy.

Plants sweat while they photosynthesize, and our textbooks teach that—like us—the hotter it gets, the more they sweat. Along the Mississippi River there are thousands of trees of the same species growing along a tidy temperature gradient: the farther south you travel, the hotter it gets. We'd developed a method to measure sweating rates by comparing the chemistry of the water in the stem with that of the water inside the leaf, which is where the sweating (or "evapotranspiration") takes place. We were startled to find that as spring progressed into summer, sweating rates went down and not up, even though the weather was getting hotter and hotter at all of the sites. It didn't make any sense to me, and the more I sweated over the problem, the more the trees did not.

We had done the field trip three times already, and my allergic reaction to the rampant poison ivy had gotten worse each time. Nonetheless, we kept anxiously wading through the waist-high fields of ivy in order to find the stubborn trees that we'd first fixed upon for sampling. I wouldn't and couldn't let the study go, and the horrible itching that I felt was nothing compared with the discomfort that I felt when each dataset looked completely different from the way we thought it should.

During our most recent trip, a rash had raged up my neck and onto my face, giving rise to a massive edema at my right temple that not only made me look like the Elephant Man but also pressed against my right ocular nerve until I lost partial vision in that eye. I knew it was really bad when Bill stopped taunting me with the nickname "Meathead" while we drove tensely back to Atlanta from Poverty Point, Louisiana (a real place), stopping only to drop me off at the Emory Hospital emergency room.

After obtaining my written consent for photographs, because "we could probably publish this," the doctors injected me with methylprednisolone and then brought in the cameras. They positioned me on tissue paper and clicked away while I tried not to giggle at the absurdity of the idea that our hopeless plant study might result in a publication after all.

After a few more hours of waiting around, I realized that I didn't have the cab fare to get home and started to wish that I had asked Bill for some cash when he had dropped me off. The Rat Hole was somewhere west of where I was; I reckoned it at about five miles away.

I lay in my hospital bed, nestled in paper, until I began to understand that I was absolutely gorgeous and fantastic. I went to the bathroom, looked at myself in the mirror, and decided that I might as well get out of there because I'd hardly be the strangest character on Ponce de Leon Avenue, especially at that time of night. By the time I got to the nurses' station, I was also pretty sure that I was the next Jesus.

They discharged me anyway, and I walked, skipped, and then ran through the Druid Hills with ideas coming to me so fast that I couldn't finish one thought before another one started up. I had to get back to the lab, because I had actually remembered something important: during my agricultural science courses I'd been taught about the delicate art of irrigation and the physics of water flow through porous soils. I remembered that for every one gram of tissue that a corn plant builds, it requires almost a liter of water. It sweats the water out to cool the biochemical machinery that turns atmosphere into sugar, and sugar into leaf. As the growing season progressed along the Mississippi, the deciduous trees must have stopped growing, having built all their new leaves back in the spring. The trees were sweating less, I realized, because the growing season was over and the system had achieved equilibrium.

Yes, as summer advanced, it was getting hotter and hotter all over the South, but the trees were already getting ready for winter: their growth rate was slowing, so they were sweating less too. These trees' activities weren't passively dictated by the temperature of our world; they were part and parcel of the goals of their world, which was focused upon making leaves. I started thinking about the American Geophysical Union conference in San Francisco—thousands of important scientists all in one place. I had to get there and spread the good gospel of my revelation.

I arrived breathlessly at the lab and eagerly revealed to Bill my crowning inspiration: we were going to go to the conference without any travel funds, personal or professional. I had figured it out—we'd drive! Sure, the meeting was in California and we lived in Georgia, but it was eight whole days away, which allowed plenty of time for us to get there.

My reasoning went like this: three-thousand-odd miles at sixty miles per hour came out to just fifty hours of driving, and we could break it into ten behind-the-wheel shifts lasting five hours each. That's five days of driving with only one shift of driving on each day, if we took two students with us. Five *easy* days of driving. We'd fill out the paperwork to rent one of the university vans that came with its own gas card, and camp all along the way (technically illegal, but still). It would then take months for the accounts to come due, and by that time I'd have the funding somehow because one of my many proposals just had to result in a contract eventually. And after all, you can't get funding if no one knows who you are, and so you have to appear at every conference and show them who you are, right?

I had previously submitted a vague abstract describing the Mississippi project to the meeting. Within it, I had hypothesized that the plants in my study used water more to support rapid growth than for cooling. It was an early attempt to shift focus away from the idea that the environment controls the plant and into a scenario where the plant controls the environment, a theme that I would reprise many times within many contexts over the coming years. But for that early conference, I didn't have a clear narrative in place. I just hoped against hope that I'd have somehow worked out what to say at the conference by the time that I got there—if I ever got there.

I began to rattle on to Bill about how the road trip is the only essentially American literary vehicle, as first pioneered by Huck Finn on the Mississippi. As with many of my manic filibusters, what I lacked in coherence I more than made up for in enthusiasm. Bill rolled his eyes. "Shut your meat-hole and go home and get some sleep," he advised. I scurried home and before long all hell broke loose as I soared up to a peak of mania and then crashed into an abyss of depression, all within the miserable privacy of my locked apartment.

Now I was back, it was days later, and Bill was looking me up and down. He brushed off the awkwardness and urged me to "snap out of it because we're hitting the road!" He was shaking our ragged Michelin Georgia state road atlas in one hand and the van keys in another. I stood amazed. While I had been mysteriously out of commission, Bill had taken my delirious suggestion as a serious order, reserved the van, and packed our gear. I smiled weakly, grateful for yet another chance to start over and do better.

The timing puzzled me, however, as it was already Wednesday and my talk was scheduled for eight o'clock on Sunday morning. We were now looking at three, not five, very long days of driving in order to get there by Saturday night. We started digging through the cardboard box where we kept our stash of the free state highway maps that we had stealthily removed from our local AAA office during multiple nervous visits staged for that very purpose. "Alabama, Mississippi, Arkansas, Oklahoma, and goddamn it—we don't have Texas," Bill bemoaned the one map out of fifty that we hadn't thought to steal. "How in the world did we forget Texas?"

"Okay, so forget Texas! Let's go north," I suggested. "Have you ever been to Kansas?" Bill shook his head. "Well, you're about to go," I assured him as I pulled Kentucky, Missouri, Kansas, and Colorado out of the box.

I laid the maps end to end and measured with my hands; if we went up and across on I-70, the halfway point was at about Denver, and I had friends in Greeley who would surely let us stay and sleep off the first all-nighter, making us ready to roll again by noon on Friday. From there it was maybe fifteen hours to Reno, where we could camp at relatively low elevation before crossing Donner Pass and coming down into San Francisco by Saturday night, where Bill confirmed that his sister was ready to welcome us into her town house for the duration of the meeting.

Sure, it was already the first week of December, but I was from Minnesota and everything would be fine. "Salt Lake City? You're going to love it," I assured Bill. "It's like an ocean of frozen mercury—it's like nothing else." I went on in raptures about the prairies, plains, and mountains we would cross, until I had fully regained my clear-eyed ambition.

"There's simply no way that this is not a good idea," I asserted, and we couldn't help but laugh at the sheer ridiculousness of this claim, while simultaneously being thoroughly convinced of its veracity.

As we packed up the van, I was further amazed to hear that Bill had indeed invited a bunch of students, and had even gotten a couple of takers. My graduate student Teri was certainly coming: she had recently returned to graduate school after working in the real world as a consultant for ten years, and I felt that it was urgent that she should begin networking as soon as possible. Though we suspected that Teri hadn't been out of Georgia much, we didn't know just how little before that trip.

Noah—our genius undergraduate who could do almost anything but never said a word while doing it—had agreed (silently, I presumed) to come. He had no driver's license and so wouldn't be of much help with the driving, but I relished the idea that he'd open up to us over the fifty-plus-hour drive and that we'd really get to know him. Bill, who had already spent a lot of time with Noah, was less optimistic about this and had already started referring to him as "the warm-blooded cargo."

We double-checked the maps and the camping gear and then drove to Chevron and filled the sixteen-seater van's obese gas tank. After that we drove to Kroger and I stuffed our large cooler with Diet Coke, ice, bread, Velveeta cheese, and baloney while Bill organized the candy trove. We agreed that the three drivers would rotate through three-hour shifts behind the wheel, allowing each driver to have six hours of rest between shifts. Thus we would stop once approximately every two hundred minutes to change drivers, buy gas, and take a bathroom break during the swap. All food was to be provided from the cooler, and whoever rode shotgun got to control the radio and was also responsible for making the driver's sandwich to his or her specifications. Bill had procured four empty two-liter bottles, individually labeled them, and placed them behind the backseat for urinary emergencies.

Our plan actually worked surprisingly well and the first twenty-four hours of driving passed uneventfully. I was behind the wheel at about midnight and merged from I-64 onto I-70 as we crossed over the Mississippi River into Missouri. Bill kneeled on the seat beside me, gazing up, his whole upper body out the window, drinking in the gorgeous Gateway Arch of Saint Louis. As we turned north and passed right under it, the full moon lit it from above, perfectly complementing the floodlights below. When Bill finally sat back down as we turned west out of Old North Saint Louis, he said thoughtfully and without even a whiff of irony, "This is a beautiful country."

I waited a minute or so and then answered, "Yes. It is." Behind us, everyone else in the van was asleep. Five hours later the sun rose at our backs and slowly opened a door of light onto the endless windswept fields of Kansas. "This is a beautiful country," Bill said again quietly to himself, and again I answered him, "Yes. It is."

We arrived at Calvin ("Cal") and Linda's home in Greeley around dinnertime the next day, piling out of the van like a pack of tired hunting dogs coming home after a long chase. I had telephoned before we left Atlanta to tell them that we were coming, but had presumed upon their surrogate parentage that I'd be welcome and could bring friends. I wasn't wrong: Cal and Linda had both been teachers for decades and loved me better than I deserved. We stretched our legs, came inside, and hungrily accepted the warm food that they offered.

"So what brings you through northern Colorado during December?" asked Cal in his offhanded way.

"We couldn't find a map for Texas," Bill answered, and I had no choice but to shrug my validation of his explanation.

Linda found a bed for each of us within their huge barn of a house and we all slept like rocks for ten hours; Bill and I were the first to emerge the next morning. Cal invited us to walk over to the neighborhood coffee shop and we accepted enthusiastically. As we drank our coffee, I told Cal about my plan to drive up to Laramie, over to Salt Lake City, then on to Reno and up over the Sierra Nevada, and finally down into Sacramento and across to the Bay Area, all in the next day and a half. Cal had grown up on a cattle ranch during the 1940s and was a steady man of few words. He listened and nodded his understanding, and then advised me in a measured tone, "There's a big storm coming in. You might want to take I-70 through Grand Junction in order to avoid the Divide."

"No, I don't, because that's longer," I said, and then boasted, "I've already calculated it."

"Well, not much longer, if it is at all," Cal replied, though good-naturedly.

Determined to win what I saw as an argument, I insisted, "No, I'll show you when we get home." We unfurled the maps of Colorado, Wyoming, and Utah and used a string to compare the two routes. With victorious satisfaction I was able to show that the string was ever so slightly shorter when it was laid up over Cheyenne. Cal just shook his head and asked us if we had chains in the van. I explained for the tenth time that no, we didn't need any of that stuff because I had grown up in Minnesota. Cal shook his head again, went outside onto the porch, and stared at the northwestern sky for a while.

Of all the regrets of my life, winning that game with the string features prominently among them. My route was indeed shorter—all of sixty miles shorter than Cal's. A single hour of driving: that's what I thought I was avoiding by hurtling us headlong into the worst winter storm of the 1990s.

As we reloaded the van, Cal and Linda's eight-year-old daughter, Olivia, decorated the inside with the crayon-colored flags that she had copied from the world atlas. While we hugged goodbye, I thought briefly about how I was always leaving the few people in the world who loved me, and then I shook it off and got behind the wheel.

My job was to drive to Rawlins, Wyoming, after which Teri would take over and get us to Evanston, just near where we would cross into Utah. As I drove, I heard Bill and Teri bickering about "the deli"—Bill's name for our cooler of food, which was by now a couple of inches of cold water with baloney floating in it and a few chunks of residual ice propping up some soggy cheese. It stank so horribly that Teri was pushing for a new rule: its lid could be opened only when two or more people wanted something inside, and even then only with the windows open. I sympathized with Teri, knowing full well that the stench was hardly going to improve in the next two days, but I felt obligated to side with Bill, as he was the only one of us who was still actually eating from the cooler. Because of the fight, everyone—maybe even Noah too—was in a sour mood when we stopped at a gas station on the west side of Rawlins so that Teri could take over the driving.

While I was waiting for everyone to finish in the bathroom, I stood and stared at the horizon, noting that the sky looked awfully dark for 1:00 p.m. I could also feel the temperature plummeting as the wind picked up. Teri came out of the gas station and climbed into the driver's seat, and I asked her to tap on the horn to signal to the others that we were ready to roll.

Bill and Noah jumped into the backseats, and Teri started the engine. I was tired and bored though the day was still young, and the map that I was holding made the road ahead look flat and uninteresting. I took off my boots and put my bare feet up against the heaters on the dashboard. I thought about putting my seat belt on but then decided against it: This was the flattest place on Earth; what could possibly happen?

As we merged onto I-80, Teri pressed forcefully on the accelerator and the van sped forward as if she were joining the commuter rat race on the Atlanta beltway. I shifted in my seat uneasily but said nothing for a mile or so. The very moment that we crossed the Great Divide, the weather changed drastically and I saw flurries starting to come down.

I looked at the wet road and realized that in the next few minutes it would become slick with ice. I looked at Teri and realized that *her* plan was just to keep ramming on the gas pedal and maintain eighty miles per hour. I spoke in the calm and steady voice that I use to instruct students while they are immersed in a dangerous and complicated laboratory procedure. "Okay, so it's going to get really icy here and you're going to want to *slow way dowwww* . . ."

I didn't get to finish my sentence, because instead of gradually slowing down, Teri had popped on the brake and, finding the road icy indeed, slammed down even harder until the brakes locked up. When we started to slide she compensated wildly with the steering wheel, and the van began waltzing in big, swinging fishtails as it flew forward. Teri was screaming at this point, having fully lost control of the vehicle, and I realized with horror that there was no way this was not going to end in some kind of crash.

The last upright thing that I saw was the speed limit sign snapping like a Popsicle stick as the van managed to hit the one single vertical object within a ten-mile radius of where we were. We spun around and around, and when we finally slowed we were facing backward into the oncoming traffic. My terror that another car would hit us was overtaken by a sickening awareness that the van was more than leaning to one side—it had begun to tip over. I tried to brace myself against the dashboard as I felt us slowly roll sideways into the ditch, accompanied by the sickening crunching sound of metal, the clattering of plastic, Teri's high-pitched shrieks, and what sounded like the fire of musket balls that heralded the first volley of the Civil War.

I was amazed at how slowly everything seemed to be happening, like a roller coaster going over the tallest peak of the ride. My head hit the cool glass of the window and then bumped and rested against the thin felt covering of the van's ceiling. All at once we were profoundly at rest. I opened my eyes and stood up awkwardly, the ceiling now serving as the floor. The other three passengers hung upside down like parachuters, effectively suspended by their seat belts.

I commenced running back and forth on the ceiling of the van, trying to check on everyone. Miraculously, we were all uninjured, save for my bloody nose, which gushed with vigor when I started to laugh hysterically. Bill was the first to unfasten his seat belt and fall ungracefully to the ceiling, and I noted that he didn't seem much fazed by the whole thing. Noah was in the far back, morosely wiping his grungy hipster hairdo with both hands. Teri simply hung there, looking dejected.

I started to worry that the van might blow up, since this is what always happens in the movies after a crash, but I wasn't sure what to do about it. Suddenly the back doors of the van flew open and a man's voice announced, "I'm a veterinarian. Is everyone okay?"

Apparently the car behind us had watched us go into the ditch, and the driver had pulled over to help.

I could hardly contain my relief; I was ready to throw my arms around the guy and kiss him. "Hey, yeah, we're fine!" I beamed.

"Weather's getting worse. Let's get you all into town." I looked past our new friend and saw a second set of Good Samaritans ease toward us, hazard lights flashing.

"Okay," I agreed happily. "Let's do it!"

The men helped us out and I was the last to leave, more because I had to dig my boots out of the van's spectacularly dispersed contents than because I was the captain of our capsized ship. We jumped into their trucks in pairs and drove off toward who knew where.

We didn't know the drivers and we didn't know where we were; we didn't have any vehicle, any money, or any real plan—and I felt awesome. I was so glad to be alive that I thought my heart would burst through my chest. I was so grateful that nobody was hurt that I wanted to sing out at the top of my lungs. Whatever came next, no matter what it was, it was going to be a gift that I could never hope to deserve. I looked back as we drove away and saw Olivia's flags fluttering across the ditch and onto the road; the yellow cross of Jamaica on its field of black and green caught my eye, and I smiled as I watched it scamper off into the distance.

Twenty minutes later we were dumped off at a filling station on Spruce Street in West Rawlins. I thanked our rescuers profusely and the more I talked, the more I could sense that they just wanted to get out of there. Teri looked positively suicidal, sulking darkly on the outskirts of the group. One of the men took Noah aside and said, "Hey, don't worry about it. You've had quite a scare." It was only then that I realized how filthy we all were. When the van had overturned, so had everything in it, including the deli. It was particularly unfortunate that one of us had failed to adequately secure the top of our two-liter bottle after use, and by the look and smell of things, I surmised that Noah had been soaked in someone's urine during the crash. I supposed that the man comforting him had assumed that the poor kid had somehow lost continence onto his own head, and I briefly contemplated setting the record straight.

Bill interrupted my thoughts. "Well, whaddya know?" he said brightly. "We've got Triple-A!" Prior to abandoning our vehicle, he had removed the gas card from the glove box and started reading its fine print. Upon hearing this news, I turned to Bill and grinned in delight. "I'll go phone and get them to yank us out of the ditch," he said, and walked off toward the pay phone.

"Tell them we're at the Super 8," I yelled up to him as I noticed the motel up the block. When Bill came back from the phone we picked up our backpacks and started walking to the motel. Once inside, we found that the lobby stank so badly that by comparison we had nothing to worry about.

I greeted the woman behind the counter. "Hi there. We're gonna stay here if you let us."

"Single room is thirty-five; double room is forty-five," she told me without looking up or removing the cigarette from her mouth.

I looked at Teri, who was clearly still in shock. "How about three rooms," I said. "Singles for them, and he and I can share," I added, indicating me and Bill. "Hundred and fifteen dollars total, right?"

"And tax," the woman added.

"And tax. You bet," I said with a smile, and timidly provided my credit card. To my surprise, she accepted it and slammed it through the manual system of pressed receipts.

"All right, this just gets better and better," I said, and then added, "Now, who wants dinner?"

Teri was sullen. "I just want to go to bed," she said, and I couldn't tell whether she was angry at me or at herself. I wanted to ask her if she was okay, but then I thought that maybe that wasn't the right thing to do, so I stood there and did nothing, which I also knew wasn't the right thing to do. Noah had disappeared as soon as his room key was in his hand, so Bill and I left the motel and walked down Elm Street looking for a restaurant. We found a greasy steakhouse, ordered two rib eyes and two Cokes, and ate with gusto, only then realizing how hungry we were.

The walk back to the motel was a lot like every other walk we'd ever taken together, and yet something had changed. We were like two mobsters who had killed the wrong guy; something about the whole near-fatal debacle had bound us together forever. We got back to the motel and let ourselves into our room. There was a king-size bed, with a grotesquely patterned burgundy quilted bedspread on it, covering sheets that definitely hadn't been changed. The dark-paneled walls and heavy polyester curtains reeked of smoke and sweet-smelling disinfectant. The carpet was stained and sticky enough that we kept our boots on.

It was late at night, and while my body was beyond exhausted, my mind was still on and glowing. Some bruises had started to make themselves known, and I had seen some blood in my urine when I went to the bathroom at the restaurant, but it hadn't upset me. On that night I felt like nothing in the world would ever happen again that would be worth getting upset about.

Bill and I lay side by side on the bed, staring up at the water-stained ceiling, dimly lit by the room's single desk lamp. The faucet in the bathroom dripped, marking a gentle, steady beat. After about twenty minutes or so, Bill said, "Well, it's finally happened. One of the students tried to kill us."

The ridiculousness of it all, put that way, made me giggle. My giggle turned into a laugh. I kept laughing, until I was laughing harder and harder from deeper and deeper inside. I laughed until my stomach cramped up and I couldn't breathe properly. I laughed until I couldn't control myself and I wet my pants just a little. I laughed until it hurt so much to laugh that I was begging not to laugh while I was laughing. I laughed until it sounded like I was crying. And Bill laughed too. We laughed out our joy and thanksgiving that we had somehow cheated Death and cheated him big-time. Our great good luck was a gift from Heaven and had revealed a world that was too sweet to leave. We would have another undeserved day and we would have it together. When our laughing finally tapered off, it was because our bodies were exhausted. I rested until I began to giggle again. Then I started to laugh, and Bill laughed too. We did it all over again. We lay side by side, fully clothed, and laughed and laughed with our boots on.

Bill got up and went into the bathroom, but came out directly, saying, "Guess what, the toilet's clogged. I knew we should have grabbed the bottles as we bailed."

"Just piss on the carpet," I suggested. "I think that's what people have been doing."

He reacted with disgust. "Don't be an animal. The bathtub drains just fine." I got up and took his suggestion, and then we both lay back down, side by side, and continued to stare at the ceiling.

"You know, I feel bad about Teri," I confessed. "She probably hates me."

"Oh, c'mon, she should be ecstatic that she's alive," Bill said adamantly.

"She should be glad *we're all alive,*" I added with emphasis, but I was troubled. "I'm sure she blames me for this mess. And ultimately, I guess it is my fault; I'm the one who signed her up for the conference in San Francisco."

"You made her travel across the country for free? In order to meet the people that she'll be trying to get a job from after she graduates? Yeah, you're a bitch all right. We should have stayed in Atlanta, where I can do all her lab work for her," said Bill, voicing a hard-edged resentment that I hadn't heard before. "She's an adult," he continued. "Shit, she's like *thirty-five* or something; that's a hell of a lot more grown-up than we are."

"Well, that's not saying much," I countered. "But it's not like anybody gave a crap if I ever went to a conference back when I was a student." I retreated into my own resentments.

"Listen, you're never going to be friends with the students, so just get that through your head right now," Bill sighed. "You and I are going to work our asses off, teach them shit over and over, risk our fucking lives for them, and they are going to unfailingly disappoint us. That's the job. That's what we both get paid for."

"You're right." I played into his cynicism, but only halfheartedly. "We don't really believe that, do we?"

"No, we don't," Bill admitted. "But tonight we do."

I lay with my eyes closed and counted the drips, soft and regular, as they fell from the bathroom faucet, until Bill finally said, "But you do know that you can never be friends with the people that you work with."

I opened my eyes because his words had unexpectedly stung me. I ventured, "What about us? I mean, we're friends, aren't we?"

"Nope," he answered, and then continued, "You and I are just two sorry sonsabitches stranded in the middle of nowhere trying to save twenty-five bucks on a hotel room. So shut up and go to sleep." And so we did, on opposite sides of the big bed, with our clothes and boots on. I decided that this must be what family feels like, and I thanked God for the day we'd had, and also for tomorrow while I was at it.

The next morning we woke up late, and by the time we emerged from our room, it was a clear, sunshiny day. Teri had been waiting for me in the lobby, fuming. It didn't look as if she'd slept at all that night.

We walked across the street to the Big Rig Truck Stop and shared a single order of bacon and eggs, which provided more than enough food for four people. After Bill ordered his eighth cup of coffee, Teri looked at me and said, "I want you to take me to the Salt Lake City airport, so I can fly back home." I nodded and got ready to tell her that I understood and that it would be no problem.

Before I could open my mouth, Bill exploded. "*What?*" He threw his silverware down and grabbed the table as if the very Earth were shaking. "You *roll the van* and now your plan is just to bail out of here and leave us to deal with it?" he asked. "*That is cold.* That shit is just fucking *cold.*" He shook his head, appalled. Teri got up hastily and left, probably to go and cry in the restroom. I considered following her and telling her that everything would be all right and that everybody makes mistakes and that the whole trip had been a stupid idea and that we'd all just go home. But my intuition as a scientist told me that it would be a mistake for us to give up that easily.

I sat at the table and thought as the dust was settling. Like everything else in the lab, the accident was ultimately my responsibility and the buck stopped with me. Last night I'd known that once morning came, I'd have to crawl out of bed and deal with the whole mess, none of which I had even remotely started to sort out. I had no idea where the van was, or my suitcase for that matter. I didn't even really know how close or how far we might be from Salt Lake City. I knew that it was now less than twenty-four hours until my presentation at the conference, and that we still had to cross three entire states to get there. But mostly I was just glad to be sufficiently alive to get to try to solve these problems. I didn't foresee anything that might kill me that day, and not getting killed was my new bar for what constituted a good day. There was nothing for it but to eat some bacon and then improvise.

Although I agreed that pushing forward was the right thing to do, Bill's reaction had surprised me. I slowly realized that he might have considered—yet never acted upon—the idea of abandoning me, and it dawned on me for the first time that he actually had the option of bailing out of his life in Georgia. Bill had taken this latest bizarre and scary episode in his usual stride and saw no way out but to put his head down and burrow out of the huge catastrophe that he'd had no part in creating. In fact, none of this had even seemed to bother him. The thing that did bother him was that someone else might consider abandoning us—the very idea enraged him in a way that all the frustrating situations we had ever encountered could not.

I finished the thought that I had started before I fell asleep: This is my life and Bill is my family. The students will come and go, they'll be what they are, some hopeful and some hopeless, but we won't get attached. This is about me and Bill and what we can do together. All the rest of it is nothing more than background noise. I released myself from the lofty, boastful, greedy expectations of academia. I wasn't going to change the world or educate a new generation or glorify an institution. It was about being in the lab and keeping it all together, body and soul. When I crawled out of that van alive, I checked my pockets and found only one currency that mattered: loyalty. I got up, paid at the register, and then held the door open for everyone as we left. "C'mon, gang, this'll get better," I told them. "It has to."

As we all walked back to the motel, I saw what looked like our van in the parking lot, but I decided that it couldn't be because it was in perfect condition. As we got closer we found that it did look perfect, if you looked at only the passenger's side. The other side was caved in like a crumpled beer can, and the driver's side mirror was nowhere to be seen, having snapped off along with one of the windshield wipers. However, none of the windows were broken, and all of the doors on the passenger's side opened and closed normally. Bill opened the van, looked inside, and commented on its "understated luxury." The deli had flown open during the crash, and the interior reeked of piss, spoiled lunchmeat, and rancid cheese. Filth was stuck to the windows on one side because the whole mess had settled and frozen while the van had lain in the ditch all the previous night.

Bill announced that our suitcases were all there and plopped himself down in the driver's seat to try the engine. He turned the key and the engine immediately roared to life and purred while it idled. I saw Bill's face break into a huge smile. "We're in business!" he shouted. I held my nose and hopped up into shotgun position while Teri and Noah crawled into the two backseats.

We returned to the highway and drove west toward Rock Springs, Wyoming, with Noah acting as the driver's side-view mirror, silently signaling when the road was open. It occurred to me that he hadn't spoken a word during the whole trip. I put my seat belt on and checked it several times to make sure that the mechanism had caught.

We pulled onto the highway and I calculated how many hours we were from San Francisco: sixteen, maybe seventeen at the most—we'd just make it. I didn't count on the blizzard that was raging in the Sierras, but that wouldn't become an issue until later. For the moment, everything looked good. Bill suddenly exclaimed, "Oh shit! We forgot to pull over and look for the side mirror." Then he added, "Oh well, we can do it on the way back."

I was stunned; I had been so focused upon just getting to San Francisco, I hadn't even thought about the fact that we were going to have to drive this broken thing all the way back across the country. I started to say something about it. As if reading my mind, Bill pointed at me and said, "Nope, I don't want to hear it. You just sit there and think about your presentation." Then he added, "After all we've been through, it had better be good."

Compared with the trip there, the five-day conference seemed uneventful, and as soon as it ended we drove back to Atlanta, this time via first I-10 and then I-20, which included Arizona, New Mexico, and two hundred miles of Texas sans map. Each day, Bill observed that the country was beautiful, and I agreed. By the time we got to Phoenix, Teri was fully herself again, and bygones were bygones.

I returned the van late in the evening after we arrived in Atlanta, shoved the keys in the rental pool's after-hours box, and walked away. Within a month every administrator at the university was absolutely furious with me. I insisted that it had been me behind the wheel, and reassured them again and again of my complete lack of remorse, arguing that I was far too grateful to be alive to find fault with whatever miracle had granted us safety. They didn't get it, and I finally stopped expecting them to. There was one person who did get it, who got all of it, and I had finally fully realized how damn lucky I was to have him along.

11

THE LITTLE TOWN OF SITKA is probably the most inviting place in Alaska. It sits on Baranof Island facing the Gulf and is kept maternally mild by the warm currents of the Pacific Ocean. The monthly average temperature never drops below freezing, making the climate benignly hospitable for the few thousand people who live there. Not much has ever happened in Sitka, except for a few days in 1867 when the whole world focused upon it briefly.

The Alaska Purchase occurred in Sitka, replete with a formal ceremony featuring both the Russian (sellers) and the American (buyers) diplomats. They were celebrating a treaty ratified by the U.S. Senate under which the United States purchased half a million square miles of new territory at a cost of two cents an acre. The total sum that changed hands—$7 million—represented an exorbitant figure to the average American facing the rubble left over from the Civil War, which had only just ended. Opinions were divided: those in favor argued that British Columbia could be strategically annexed as a next step; those opposed despaired that the acquisition merely burdened America with more unpopulated territory to fill. In post–Civil War America, the treaty also served as escapist drama, yet another battle between good and evil, but this time conducted in a strange land, far away.

A second great drama unfolded in Sitka in the 1980s, but it wasn't a treaty between nations; it was a war between species.

Trees love Sitka. The long, light summers coupled with the mild climate make Baranof Island a darn fine place to live and grow, even if the cold, dark winters prevent the plants from growing very large.

There's a Sitka spruce, a Sitka alder, a Sitka ash, and a Sitka willow, all first identified during the exploration of the region. These Sitka trees have successfully colonized British Columbia, along with the states of Washington, Oregon, and California. And yet they are modest trees: the Sitka willow, in particular, is not an imposing plant. It reaches a maximum height of twenty-three feet—not exactly a forest giant. But with the Sitka willow, as with all plants, there is much, much more than meets the eye.

Whenever you stroll through a eucalyptus grove you are engulfed in a unique smell, acrid and spicy and a little bit soapy too. What you are actually sensing is an airborne chemical that is created and released by the trees, a "volatile organic compound," or "VOC" for short. The VOCs were synthesized to be "secondary" compounds—this means that they don't provide any nutrition, and so they are secondary to basic life functions. VOCs have many uses that we do understand, and probably a host of others that we do not. The eucalyptus releases VOCs as part of an antiseptic that will keep its leaves and bark healthy if it is wounded, preventing infection.

Most VOC compounds don't contain nitrogen, thus they are relatively cheap for the plant to produce and are therefore expendable. There's no real downside for a tree to pump VOCs through the forest in lavish excess, giving rise to that characteristic eucalyptus smell recognizable to a human nose. In contrast, the vast majority of VOCs that trees produce cannot be detected by the human nose, which is just fine because that is not what they are for. The production of VOCs within a forest waxes and wanes because individual VOC compounds can be turned on and off with a signal. One common signal is jasmonic acid, which is produced copiously when a plant is wounded.

In the war between plants and insects that has been raging for four hundred million years, both sides have had their casualties. In 1977, the state university's research forest in King County, Washington, was utterly ravaged by an insect attack. Tent caterpillars led the charge: they are brutal and insatiable warriors and were able to completely defoliate several entire trees and fatally damage many more before they triggered a crash in the local tree population across many broadleaf species. We all know that it is possible to lose a battle and still win the war, and nowhere do we find this to be more true than in the history of trees.

In 1979, back in the laboratory at the University of Washington, researchers fed leaves from the trees that had survived the attack to tent caterpillars and then carefully watched them eat. They observed that these caterpillars grew much more slowly and sickly than caterpillars generally do, and they certainly did not grow as well as they had grown on the same trees only two years earlier. Simply put, there was some chemical in the leaves that was making them sick.

The really exciting thing, however, was that the healthy Sitka willow trees located a full mile away—trees that had never been attacked—were equally unpalatable to the tent caterpillars. Indeed, when fed the leaves of the distant and healthy trees, the caterpillars became just as feeble and sickly and were rendered incapable of destroying a forest, as they had done so easily over the hill just two years previously.

The scientists knew about root-to-root signaling between adjacent trees via underground secretions, but the two groups of Sitka willow were too far separated for any soil-talk to have taken place. No, some aboveground signal must have been transmitted and received. The scientists concluded that when the leaves were first wounded, the plant began to load them with caterpillar poison, which also triggered VOC production. They further hypothesized that the VOC must have traveled at least a mile and was sensed as a distress signal by the other trees, which then preemptively fortified their leaves with caterpillar poison. Through the 1980s, generation after generation of caterpillars died miserable, starving deaths due to these poisons. By playing this long game, the trees ultimately turned the tide of the war.

Given their years of observations, the researchers were convinced that aboveground signaling among the trees was the most likely explanation. They knew that trees aren't people and that they don't have feelings. For us. They don't care about us. But maybe they care about each other. Maybe during a crisis the trees take care of each other. The Sitka willow experiment was a beautiful, brilliant piece of work that changed everything. There was only one problem: it took more than twenty years for anyone to believe it.

12

I COULD FALL ASLEEP, but I couldn't stay asleep. Off and on, for several weeks during the early spring of 1999, I'd awaken at about two-thirty in the morning and become increasingly agitated over my inability to get back to sleep. Bill ran the lab beautifully, and each experiment succeeded like clockwork, which made it all the more frustrating when each of my proposals for grant contracts was rejected, one after the other. In order for a contract to be approved, it must pass a stringent peer review. Evaluation is heavily slanted toward "track record"—the number of significant discoveries that have resulted from previous contracts; thus a brand-new researcher is at a serious disadvantage.

It is also not uncommon for scientists to work out their personal issues under the guise of making an evaluation, and I was receiving feedback along the lines of "this reviewer is dismayed to find that the investigator's apparent capabilities were deemed sufficient to merit a graduate degree from the very same institution that produced his own credentials," and other useless venom. During the San Francisco conference that we had nearly killed ourselves getting to, and at which I had presented my ideas about plants and water uptake, an enraged senior scientist (who years later betrayed himself to be a nice person) had stood on a folding chair and yelled "I can't believe you are saying this!" while I tried to speak. In my shock and confusion I had stuttered, "Something is wrong with you," into the microphone, and it hadn't rendered the atmosphere more congenial.

To be fair, the trouble had really started years earlier. While taking a break from writing my dissertation, I visited a new professor whose recent arrival I had long anticipated because of her singular expertise in paleobotany. I helped her to unpack her large collection of fossils and sort, label, and store them. The rocks contained traces of the Earth's earliest flowers and she had collected them at great personal risk within the jungles outside of Bogotá, Colombia. These sediments were 120 million years old, and my colleague planned to extract the tiny pollen grains and fern spores that had collected underneath the fossilized petals. After examining them under the microscope, she would meticulously describe the shape of each grain she found and keep a record of how the number of grains changed between rocks. Using the statistics generated, she could then discern how the appearance of flowers had been related to changes in fern populations, and measure the amount of shade that was present in the dark understory that had fomented a botanical revolution.

The rock samples were jagged and crumbly, and so dark that I wondered aloud if they might still contain enough organic carbon to be measureable on a mass spectrometer. I ran some test samples and found more than enough—in fact, there was even enough carbon to perform a new type of chemical analysis, a technique that would measure the ratio of heavy carbon atoms to more normal, everyday carbon nuclei.

Our work turned out to be some of the first analyses of carbon-13 within ancient terrestrial rocks, and though I was able to finish the lab work in less than two years, it ended up taking me six full years to interpret the data and finally publish my findings. Thus my early years as a professor were spent trying to persuade the world that I had used an unusual method on unorthodox samples to gain a surprising result via an untested interpretation. The whole thing had come out of left field, and I was naive in thinking that I could win over audiences who had decades more research credibility than I did. My early career had all the makings of a long, slow academic train wreck.

As I spent those early years repeatedly smashing against a brick wall of scholarly skepticism, my bewilderment ripened into the realization that it would take me many conferences, much correspondence, and a great deal of intellectual soul-searching to successfully convince a critical mass of other scientists that I knew what I was doing. The trouble was, I didn't have years. After I ran out of the money that the university had allotted for the start-up of my lab, we began appropriating chemicals, gloves, test tubes, and anything that wasn't nailed down from the dusty abandoned basement of the building in order to keep working. My transparent justifications to the tune of "at least I am using it for something" had begun to ring hollow as our desperation drove us to scour the Dumpsters, recycling bins, and finally the teaching labs within the engineering buildings, where they seemed to be so rich in everything, we told ourselves, that they couldn't possibly miss this or that.

The money to pay Bill's salary was the last to run out, and though he always made a great show of indignation and moral offense when a student finally worked up the nerve to ask him whether he was "that guy who lives in the building," the whole situation was beginning to wear on both of us. At first, Bill regarded his destitution as a novel adventure—a temporary bohemian phase—but it lost its meager charms as the months dragged on. Throughout the time he was homeless, my small gestures, such as cooking him dinner every night, were enough to offset much of my guilt, but lately it had become clear that I was ruining both of our lives.

I was also existentially terrified. Ever since I was a little girl, I had longed to be a real scientist, and so soon after finally getting close, I was in danger of losing it all. I was working extra hours, but the inefficient all-nighters weren't helping much.

While investigating a light that he assumed had been carelessly left on, the night custodian muttered upon finding me, "However much you love your job, it ain't gonna love you back," and shook his head in pity as he closed my office door. I didn't want to agree, but I was beginning to see his point.

The nightmare of losing the lab was all the more horrifying because it had been my only concrete dream. During my college years I had latched on to the idea that once I was a bona fide scholar (the main manifestation of that idea being a laboratory with my name on the door), everyone would acknowledge my credibility, some scientific breakthrough would logically follow, and life would be easy. I had raced through graduate school secure in my expectation of this reward.

Thus I was bewildered by my failure during these early years of professorship and deeply worried, for the first time, that my cosmic destiny would go unfulfilled and the faith of my frustrated foremothers—whom I always pictured scrubbing bedclothes while up to their elbows in lye—had been wasted on me. During the anxious melodrama of those sleepless nights, I'd start thinking about Saint Stephen—the poor bastard. How he started out all full of piss, vinegar, and the Holy Spirit but hadn't even gotten out of Jerusalem before his audience strung him up on the outskirts of town. Only a few days before that, Stephen was picked as one of the lucky seven who were expected to go forward and speak the gospel. Did they explain to Stephen that there was a very real chance that he'd enrage people with his great new take on things? Of course he was supremely pious and all that, but didn't he feel like a bit of a sucker nonetheless?

The Bible is always short on details. Didn't Stephen's innate sense of self-preservation thwart his martyrdom at all? When someone throws a stone at your head, don't you instinctively dodge it? Put your arms up? Or do you close your eyes and let it happen, waiting for a good hard smack to the temple? And where did the stones come from anyway, when they stoned people? Did people collect them on the way to the scene? About how many stones did each thrower figure that he needed? Did they examine each one they picked up, discarding and retaining them based on some criterion? Did women get to throw too, or did they just simper on the periphery, as depicted by Raphael? I thought about Saul, the elder who oversaw the whole grisly affair, and how he eventually came around to Stephen's way of thinking and gained no small amount of celebrity roaming the empire and espousing it, but only after Stephen was good and dead.

With my mind thus running in pointless circles, I'd get more nervous and then unbearably achy, starting with my knees and elbows and spreading to my ankles and shoulders. I'd sit on the edge of my bed, kneading my joints and rocking back and forth for a half hour or so, and when I couldn't stand it any longer, I'd call Bill. The ancient phone on the wall of the office where he slept had a ring not unlike an old fire alarm, and he'd answer quickly, more out of an urge to silence it than out of concern for me.

"Is it the witching hour already?" he said when he picked up the phone.

"I don't feel so good," I muttered in a shaky voice that betrayed my cascading anxiety.

"You sound like shit. Have you eaten anything since you dropped me off?"

"I drank some Ensure," I offered, and he sighed in exasperation. There was a long pause.

After a while he groaned and said, "I guess this is the part where I tell you that it is all going to be fine."

I was trying not to break down. "But what if it isn't? What if I never get a grant? What if I'm just not smart enough? What if we lose everything?" I rambled in agitation.

"'*What if*'? Fuck 'what if.' None of that shit will change anything!" Bill hollered. "What if you don't get a grant? You can't exactly pay me *less* than you do now, in case you haven't run the numbers lately. What if you get fired? We've got the goddamn keys to the place; I'll go make copies tomorrow. I've developed a nagging suspicion that you don't have to be employed here to come in and work every day. You just keep putting on that power suit and hawk our wares at those interviews, and get us *out of here* for God's sake. If we built this shit once, we can build it twice. Or we'll roll up the goddamn tent during the night and just disappear—you can grind the organ in the next town over while I run around tipping my bellboy's hat and rattling coins in a tin cup."

By then I was laughing weakly, soothed by his admonitions. There was a long silence.

"Shall we have a reading from *The Book of Marcie*?" I offered.

"Finally you begin to talk sense," Bill assented, and I dug the thick volume out from under my bed and flipped it open to a random page.

We had nicknamed one of my recent master's students "Marcie" after her own favorite *Peanuts* character. In the end, however, she proved to be more like Peppermint Patty, tending toward the same good-natured resignation with respect to her multiple D-minuses. She had recently left the lab on decent terms, having decided against improving her work enough to make it passable. Her parting gift to us was the draft of a "thesis" that had swollen grotesquely with each revision, and I kept arguing that it heralded an emergent literary style. Everything about it was ridiculous, from its fourteen-point Palatino font to the unfortunate fact that some of the pages had been shuffled in upside down prior to binding. While we waited out my insomnia, I read a three-page paragraph of Marcie's nonsense and followed it with a section of *Finnegans Wake*. Then I asked Bill to identify which one was which and to justify his determination through critical analysis. The night before, I had compared and contrasted the "Methods" section of *The Book of Marcie* with the famous "Lucky's Think" monologue from *Waiting for Godot*.

Anticipating the singular catharsis that comes of conspiring in something despicable, Bill and I provoked each other with feigned erudition. Lately these long, bantering phone conversations with Bill had become the only thing that could harness my racing thoughts so that I might sleep.

A break in our conversation led to a long pause, and when I looked out the window I could see no indication of a sunrise. I checked the clock and said, "Wow, four a.m. and I think we're there. New record." My anxiety had subsided.

"You know what's the worst part of this for me? It's that I am sure you are keeping the Beast up, damn it," lamented Bill. I looked over at Reba, who was indeed lying in her basket at the foot of my bed, quietly awake and watchful.

Another long pause followed. "Cripes, why don't you go to a doctor or something?" Bill asked me, in a voice that was almost tender.

I laughed off his suggestion. "No money, no time, and for what?" I answered. "So he can advise me to reduce my stress level?"

"So he can prescribe you some fucking Prozac."

"I . . . I don't need it," I said.

Bill's answer came quickly. "Then don't take it," he said. "Give it to the homeless guy who lives in your laboratory."

A new wave of guilt crashed over me as I realized that this was as close as Bill had ever come to admitting to me that he was unhappy.

"I'll think about it," I promised. I put my hand over the mouthpiece so Bill wouldn't hear me choke back what I wanted to say. Finally I gave voice to part of it and said softly, "Thank you for picking up when I called."

"That's why you pay me the big bucks," said Bill, and then he hung up.

* * *

Things would get better. Six months later, we rented a moving van, packed it with scientific equipment, loaded Reba into the front seat, fastened our seat belts for a change, and drove north toward Baltimore. I had secured new jobs for both of us at Johns Hopkins and convinced the two universities that it made more sense to simply transfer laboratory instruments than to discard them. After we moved, I took Bill's advice and went to the doctor. I got on the right medications and started eating healthy and sleeping regularly, and I got stronger. Bill quit smoking. We both kept working, kept pounding on doors, and kept believing that eventually they would have to start opening.

Love and learning are similar in that they can never be wasted. I left Atlanta knowing more than I had when I arrived. To this day, I need only close my eyes to summon the smell of a crushed sweet gum leaf, as pungent as if I were holding it in my hand. Point to any object in my laboratory, and I can tell you how much I paid for it down to the penny, and which company sells the cheapest version. I can explain the theory of hydraulic lift so that every single student in the room understands it on the first go-round. I know that there is more deuterium in soilwater from Louisiana than there is in soilwater from Mississippi, although I am only about halfway done figuring out why. And because I know the transcendent value of loyalty, I've been to places that a person can't get to any other way.

Part Three

FLOWERS AND FRUIT

1

FOR SEVERAL BILLION YEARS, the whole of the Earth's land surface was completely barren. Even after life had richly populated the oceans, there is no clear evidence for any life on land. While herds of trilobites wallowed on the ocean floor, preyed upon by *Anomalocaris*—a segmented marine insect the size of a Labrador retriever—there was nothing on land. Sponges, mollusks, snails, corals, and exotic crinoids maneuvered through nearshore and deepwater environments—still nothing. The first jawed and jawless fishes appeared and radiated into the bony forms we know today—still nothing.

Sixty million more years passed before there was life on land that constituted any more than a few single cells stuck together within the cracks of a rock. Once the first plant did somehow make its way onto land, however, it took only a few million years for all of the continents to turn green, first with wetlands and then with forests.

Three billion years of evolution have produced only one life form that can reverse this process and make our planet significantly less green. Urbanization is decolonizing the surfaces that plants painstakingly colonized four hundred million years ago, turning them back into hard and barren lands. The amount of urban area in the United States is expected to double during the next forty years, displacing a total area of protected forest the size of Pennsylvania. Within the developing world, urbanization is happening even faster and involves much more space and many more people. On the continent of Africa, a Pennsylvania-sized area of forest is converted to city every five years.

Baltimore is the most tree-impoverished city on the East Coast of the United States, where the relatively wet climate used to support dense forests. Baltimore proper contains about one tree for every five of the city's inhabitants. Viewed from space, only about 30 percent of Baltimore City appears at all green and the rest is uninterrupted asphalt. On the same day that Bill and I arrived in Baltimore, I got a mortgage with no down payment and bought an old row house near the university. Bill moved into the attic and quickly became reaccustomed to not sleeping in a public building. It was bittersweet to have left Georgia, a place where we had both grown up so much. But like the very first plants, we needed a new place to spread out, and so we decided that this bare slab of rock could serve as home.

2

"DO YOU REALLY THINK this is illegal?" I asked Bill over the CB radio.

"Jesus, I don't know. Let's mull it over using the public airwaves." Bill's voice was crystal clear, which was unsurprising given the fact that he was driving the vehicle directly in front of me. We were returning to our still-new home in Baltimore after a quick trip to Cincinnati, and one of us was driving a U-Haul van.

"Well, I'm just thinking," I mused. "We've got at least four hundred miles to go, and if a cop pulls us over and notices that we're hauling hundreds of dollars of lab equipment stamped 'Property of the University of Cincinnati,' our Maryland state driver's licenses may not qualify as sufficient proof of ownership."

"Don't you have a copy of Ed's last will and testament, stating, 'Being of retired mind and body, I hereby bequeath all my laboratory possessions, contaminated and otherwise, to my academic granddaughter that she may go forth and multiply my own findings manyfold'?" Bill was actually perfectly happy to converse, and I had been put under strict orders to keep babbling after the radio in his vehicle had proved inoperable.

"No, I don't have one, and I don't think that he'd want to put anything in writing anyway," I said, considering. "Maybe I'm just being paranoid," I added. "I mean, what possible crime could a cop think we were planning to commit with a bunch of beakers?"

"I dunno, you dumbass, how about inaugurating the ten-thousandth methamphetamine lab in West Virginia, for starters?" Bill impressed me with his worldliness.

I didn't think that he was being very supportive, seeing as he had been at least as keen to take the stuff as I had. Hadn't he packed and repacked the U-Haul three times over, successfully cramming in more boxes during each reconfiguration?

"Listen, you're right; we need to keep our eyes on the prize here, and remember that all this useless crap was *free,*" he said, morally clarifying the situation.

The next day we'd continue our work converting a huge basement room within the Johns Hopkins geology department into a magnificent laboratory, a project that we started after we moved during the summer of 1999. In between major construction tasks, we had been making the rounds at all the national scholarly conferences on biology, ecology, geology, whatever—I was getting my name out there and generally promoting the new lab. While wandering through the vendors' hall at the Geological Society of America conference in Denver during the fall of 1999, we had bumped into my favorite "academic uncle," Ed, while he was busy searching for his wife's birthday gift. I hadn't seen him in some time; he was a little grayer but still cut the same fatherly figure with which I associated him. When I approached him to say hello, he had stopped what he was doing and greeted me with a big hug.

Ed had gone to graduate school with my dissertation advisor (hence the "uncle" designation) and had been one of the scientists who figured out the ups and downs of sea level over the eons. He and his team had analyzed thousands upon thousands of tiny ocean shells left by the microscopic animals that lived and died at the surface of the ocean. This work had started during the sixties and led to a method by which the chemistry of a shell could be used to calculate how much ice was sitting at the North Pole, based on a series of serendipitous indirect relationships.

When Arctic summers are cold, the snow that falls during the winter does not melt but instead builds up and packs down on itself until huge tongues of ice are forced out from the bottom of the pile. We find the tracks of such splaying ice as far south as Illinois, leading us to argue over whether unremittingly cold summers could create a "snowball" Earth, covered in ice from pole to pole. Because precipitation starts as evaporation, an Earth with huge expanses of polar ice is also an Earth with less ocean water, enough to lower sea level by many feet. With the sea thus drawn down, new land is exposed, and a new type of real estate opens up for plants, animals, and people. Bodies of water that have kept animals separate for thousands of years dry up, and everything begins to mix. An icy world is a brave new world, full of land to conquer and balances of power to be challenged.

Ed and his contemporaries dared to believe that this cooling and warming occurred in cycles, but mourned the fact that each generation of ice would have rubbed out the tracks of the one before it, forcing them to search for new ways to read history back beyond the most recent endless winter. At the bottom of our swelling and retracting ocean, the empty shells of tiny organisms that lived their brief lives at its surface collect, and the drills that search for petroleum haul up layers and layers of the rock into which these shells have solidified.

Each little shell bathed in the ocean of the time when it lived, in the water that was left over after ice had been taken out. During this bathing, the chemistry of the ocean was imprinted upon the chemistry of the shell, leading to the theory that analyses of fossil shells through time tell the history of global ice—of the glacial cycles. For decades, Ed had worked on his little piece of the theory, which progressed from being an unlikely fancy to a reified fact and is now found within every introductory geology textbook. To do his work, Ed ran a large laboratory full of state-of-the-art equipment, that "state" being 1970.

Ed asked me what I was up to these days, and I told him that I was building a new lab at Johns Hopkins University. I introduced him to Bill, whom he didn't quite remember from Berkeley. I knew that since I'd last seen Ed, he had been promoted to dean, and I asked him how he liked it. "I don't," he told me while picking through a tray of gemstones. "I'm retiring at the end of this year."

Although he was easily in his seventies, I was nevertheless shocked by this announcement, as I was far from ready to think about losing the generation that had mentored me. I wondered what the Old Boys'

Club would make of me once my few allies, such as Ed, were no longer around to defend me behind closed doors.

"What's going to happen to your lab?" I asked Ed in disbelief.

"It will house a computer cluster for the new geophysicist that they just hired," he told me sadly. "All my stuff will go straight into the Dumpster. Why, do you want any of it?"

The blood rushed to my head. I looked at Bill, whose mouth was hanging slightly open. The next week, we got in my car and drove to Ohio. When we got to Cincinnati, we rented a U-Haul for the one-way trip back.

When we met Ed in front of the building that housed his laboratory, it was mid-morning on a Tuesday. He took us inside and introduced us all around, proudly telling people that he'd known me since I was a new student, that I was now a professor doing great things, and that I had come to town because his equipment was of too much scientific value to be discarded. He told them the stories that I had heard him repeat every time he saw me and that he probably repeated when I wasn't around too. He told about how I had written him a long letter after reading one of his papers, asking for the backstory of the experiments, for the details, and for the "blooper reel." He told about how he had gone on a soils trip with me and I had slept in my car because I didn't want to lose valuable daylight hours setting up a tent. He told them that I was the hardest-working student he'd ever seen and that he knew I was special from the first time that he met me. I kept my head down so that no one could see my embarrassed smile and experimented with standing on one foot while waiting for him to finish.

After he finished, I looked up at Ed and said, "Thank you." Then I cringed as, one by one, the people to whom I was being introduced sized me up and down, each of them wearing a look with which I was very familiar. It was the look that says, "*Her?* That can't be right; there's a mistake here somewhere." Public and private organizations all over the world have studied the mechanics of sexism within science and have concluded that they are complex and multifactorial. In my own small experience, sexism has been something very simple:

the cumulative weight of constantly being told that you can't possibly be what you are.

"And you don't do yourself any damn favors by going around in pigtails and stained T-shirts," Bill reminds me whenever I affect an air of persecution, and I must concede his point.

Ed took us down into the basement and unlocked his lab for us. It was clear that no experiments had been done inside that room for years, but nonetheless it constituted about a thousand square feet full of dusty instruments as well as a sizeable hoard of supplies. Bill was standing in one corner, and the look on his face told me that he was mentally comparing the volume of the room with the volume of the truck. I knew that his first instinct would be to simply transfer all objects from one to the other and that I would have to make a stringent case for leaving anything behind, up to and including a drawerful of used earplugs.

"Well, you'd better let us know what you would miss if we took it." My mind was bucking with greed, making it difficult for me to sound diplomatic.

Ed smiled. "You know, I couldn't even tell you what most of this stuff is anymore. The guy who worked for me—a brilliant guy—his name was Henrik—too bad you never met him—he custom-made most of the instruments. We worked together for thirty years. He retired three years ago and lives in Chicago now, but you could contact him if you needed help figuring something out. Even the factory-bought stuff he had to adapt quite a bit. He only had one arm."

A long silence followed, which was broken only when Bill raised both hands to the ceiling and bellowed out, "Dear God, do you mean he was a *gimp*? If there is *one* thing I cannot tolerate, it is the idea of a freak in the lab! Disgusting!"

During the strange minutes that followed, Ed turned to look at me as if to say, "Where on Earth did you *get* that guy?" I just stood there paralyzed with a serene smile on my face, as I usually did on such occasions. Ed shook his head and checked his watch, saying, "I should get back to the dean's office. The guys in Facilities might help you with anything heavy if you ask. Come by my office when you are loaded up—the secretary upstairs will tell you where to find it." He pulled a necktie out of his briefcase, donned his suit jacket, and walked out.

I grinned at Bill as we locked eyes. "I can't take you anywhere," I sighed.

Bill is missing part of his right hand, which is also his dominant hand. For some reason it takes people years of working in close quarters with him to notice it, if they do at all. Because his skin is extensively scarred, it is clear that his hand started out whole and that a hunk got chopped off at some point. He must have been a very young child at the time, because Bill has no memory of the event. I think the only people who truly know what happened are Bill's parents, and they aren't at all interested in talking about it. Bill's mother has Swedish blood, and so all this lack of information makes perfect sense to me.

Bill can do more with 1.7 hands than the vast majority of people in the world can do with two, and so the only time that the nonstandard nature of his appendage matters is when it provides the odd comic moment. I experience twisted glee when I insinuate to people that Bill received the injury during a bungled lab experiment, and one of Bill's favorite pastimes involves walking up behind a student who is working with a sharp scalpel and barking out, "Watch yer digits!"

We carried in the cardboard boxes and rolls of bubble wrap that we had brought with us, and moved furniture in order to set up a staging area where we could pack the items we wanted to take. We decided that Bill would disassemble the big stuff and I would sort the little stuff, wrap everything, and pack it in cardboard. We worked for hours, focusing first on items that would be of obvious use: unopened boxes of gloves, custom-sized flasks, independent transformers, pumps, and power supplies. We then moved to seldom-used but expensive items, such as containers that could slow down the boiling of super-cold fluids when they were exposed to air. For each item that I packed I imagined the several hundred dollars that we'd never need to spend, and I kept a running tab. Bill carefully drew the larger items in his notebook and photographed them from all angles before taking them apart, knowing that he'd have no other assembly manual once we returned home. He also multitasked impressively by criticizing everything I did while he worked.

"What the hell? You're using up all the bubble wrap. Slow down," Bill ordered me.

"Oh, sorry," I answered. "Stupid me, I thought I remembered something from my Ph.D. about how glass can break. I guess you learned better at community college."

"Use less wrap, it'll go further and pack tighter and we can take more stuff," he snarled in return. "I'll drive slow."

"Why are you in such a foul mood?" I asked him. "You should be glad I set up this little heist."

"Oh, I don't know," he answered. "Maybe because I drove all through the fucking night while you slept."

"Did I forget to thank you for that?" I chirped with wide eyes. "Oh well, too late now; what's done is done."

We were avoiding what we wrongly assumed would be a tense confrontation over the homemade mass spectrometer on the other side of the room. We both wanted it, but we knew we couldn't take it. Eventually we drew near it and walked around it to look at it from all angles, as if we were slowly circling wily prey. It was a big, standalone thing, about as big as a small car, fronted by a panel of analog readouts, each with a needle that had long since ceased to dance. "That thing is half glass, half metal, and half particleboard," joked Bill as our eyes tried to identify the path from the inlet to the detector through the wires, gauges, and handwritten signs that said things like DO NOT OVER-TIGHTEN ME on the exterior of the machine.

I often compare my mass spectrometer to a bathroom scale. Both instruments can be used to measure the mass of an object and report the result according to its place upon a spectrum. On your bathroom scale, the extremes of this spectrum might range from twenty-five pounds to two hundred fifty pounds. When a person steps on your bathroom scale, a spring is mechanically compressed, and the force is transferred to a dial that turns underneath a needle. Numbers have been painted on the dial to increase with increasing force.

A bathroom scale can very accurately tell you whether the object on it weighs about fifty pounds or is really closer to two hundred pounds. Your bathroom scale is great for allowing you to determine the difference between an adult and a child, but it's just not accurate enough to help you figure out the amount of postage needed for your Christmas letter. For that problem, you should use the scale at the post office, which is perhaps a tipping bar that balances perfectly when you slide a weight to the marked position where it exactly offsets the weight of the letter set on the tray.

The bathroom and post office scales are two machines, each one cleverly designed to yield the same type of measurement, the same end through different means. We can keep focusing in on that spectrum: Let's say we want to weigh two sets of atoms, and we'd like to be able to see which set is heavier due to its random incorporation of a handful of extra neutrons. We need to build a machine. The good news is that we only need to build this machine once, since there's no chance that anybody besides us will ever want this thing in their bathroom or government office. This frees us to make it as ugly, silly, unwieldy, and inefficient as we want to—we just need to improvise something that works for us. This is how scientific research instruments are built.

The creative process born from these necessities gives rise to delightfully quirky creations, unique as their creators. Like all art, they are a product of their period and an attempt to address the issues of their age. Also like art, they appear outmoded and antiquated when viewed from within the future that they helped create. Yet there is a singular fascination to be indulged when we stop and stare at the piecework of previous scientists' hands, amazed over the care taken with the peripheral elements, just as we are dazzled by the hundred tiny brushstrokes that magically agglomerate into one small boat on the horizon within a pointillistic painting.

Fifty years ago, scientists such as Ed sculpted their works around huge magnets, which served as the pulsating heart of the eventual machine. The electromagnetic field generated by any magnet acts in proportion to its mass; thus a big magnet creates a field strong enough to pull noticeably differently on different atoms. Their idea then became to accelerate two sets of atoms past the same magnet and then measure how much each was thrown off course as it flew through the electromagnetic field, and from the flight paths determine which one contained a greater proportion of neutrons.

Simple calculations showed how this should work, as the mass-dependent effects of a magnet have been known for hundreds of years. The practical problem of accelerating the particles and measuring the deflection—of actually *doing* it—was worked out by a fairly limited group of researchers working at the University of Chicago, whose students went on to improve the methods at the California Institute of Technology. Their techniques eventually spread to places like Cincinnati and many years later have been automated into the easy-to-drive versions that we use in my laboratory.

Back in those early days, as now, the sample was introduced for measurement as a gas and then ionized before acceleration. Magnetic deflection of the particle beam spewed the sample against a target, and each hit produced a tiny electrical signal. A row of detectors collected these electrical impulses and positioned them on a spectrum whose peaks corresponded to mass. Like a bathroom scale, these mass spectrometers had to be calibrated against familiar items with standard weight, but then they could be used for almost anything that could be coaxed into a gaseous state, including the shells at the bottom of the ocean.

The instrument we were looking at—Ed's old mass spectrometer—resembled a high-tech scrap metal heap and probably weighed at least a ton. Before it was loaded with sample, its metal foyer had to be mechanically pumped free of atmosphere, as did the flight tube. In Ed's day the pumps were little more than a motorcycle engine housed in a steel box, turning rapidly enough to create a strong suction that could be sustained as long as power could be delivered and the noise could be tolerated.

The gas moved through the inlet similar to the way that a barge moves through the locks of a dam, sitting somewhere until the next chamber was sufficiently pumped free of atmosphere. To seal the gas into these waiting chambers, liquid mercury flowed in, provided a wall, and then drained out when the wall was no longer needed. The metallic fluid was almost perfect—chemically unreactive, incompressible, and electrically conductive. There was the small matter, however, of it also being monstrously toxic. Bill and I stared at the beautiful old instrument, knowing that we had no use for it and shaking our heads at its glass saddlebags filled with liters and liters of shiny mercury.

A single drop of mercury from inside an old-school thermometer requires full-on hazardous-materials disposal if broken open. The mere sight of gallon jugs of mercury filled us with awe and our minds bent over the risks that must have been taken while Ed (or, rather, the brilliant Henrik, observed Bill) worked with these substances for decades. A blood-pressure cuff had been modified and added as a means by which to coax the mercury forward and backward, and could presumably be operated with only one hand. The paint on certain knobs was worn from years of careful turning, and the soldering evidenced repeated amateur attempts that finally resulted in too-strong seams. The machine itself provided unsolicited paternal advice to the user, such as "Is H2 off?" and "Turn this one LAST," written on the valves in red and black permanent marker. A bow of red yarn was tied in one odd corner, maybe to reinforce the memory of a forgettable but necessary step, or perhaps simply as a good luck charm.

After we had stared at the machine from every angle, I observed, "It's a shame to throw this out. Someone should put it in a museum somewhere."

"No one will," said Bill.

While we were walking away, I noticed something propped up behind the instrument. It was a one-foot-square piece of wood, with the sharp ends of ten or so screws sticking out of it. Their pointy tips were arranged in a grid, and under each one of them was written the diameter of the screw: one-sixteenth, three-eighths, five-eighths, nine-sixteenths, and so forth. It served a supremely useful purpose, allowing one to quickly assess the size of a stray nut, washer, or bolt, thus helping to diagnose what the hardware had fallen off of or for what it could be used.

"No wonder Ed is in the National Academy," I said. "We have to take that."

"No," said Bill. "It stays here." I was surprised by his adamancy.

"Are you nuts? It's small and we don't even need to wrap it," I pleaded.

Bill was looking at it thoughtfully. "No. It's theirs. It needs to stay with Ed."

"But it's pure genius," I argued. "It has the power to transform Western civilization and you know it."

"Relax, I'll make you one," said Bill. "I promise."

When we were finished loading the truck, we went to find Ed's office and I knocked on his door. When he opened it, I handed him four sheets of paper and said, "I made a list of what we took, just so you have it." Ed followed us outside, looked into the truck, and helped us secure everything a second time, and then it was time to go.

"Thank you for everything. It means a lot," I said, wanting to add something significant but not knowing what more to say. "You've probably even bought me a couple more years before they fire me," I added with a smile.

"Oh, I have a feeling you are going to be fine." Ed laughed as he shook his head. "Make sure not to wear yourself out along the way, okay?"

His oblique recognition of my years of effort amplified the poignancy of the situation, and suddenly a big lump gathered in my throat. There in the parking lot, we two scientists conducted a homely ceremony that transferred the tools of his life—his career—to mine.

Ed's suggestion that the Earth's ocean chemistry could be reset completely was a dangerous idea when he was young, and he had stayed up nights to study while the people he knew were watching Joe DiMaggio and arguing about the McCarthy trials. Forty years later his idea was one that I could take for granted as I dared my way into my own ambiguous future. It was kind of tragic, I reflected, that we all spent our lives working but never really got good at our work, or even finished it. The purpose instead was for me to stand on the rock that he had thrown into the rushing river, bend and claw another rock from the bottom, and then cast it down a bit further and hope it would be a useful next step for some person with whom Providence might allow me to cross paths. Until then I would keep our beakers, thermometers, and electrodes in my care, hoping against hope that not all of it would be garbage upon my own retirement.

While absorbed in these thoughts I looked at Ed and was suddenly overcome by an irrational fear that he might die before I saw him again, and I hugged him hard. I couldn't bring myself to watch as Ed shook Bill's right hand goodbye, but I did notice that their handshake had morphed into a bear hug by the time that I had gotten myself into my car and settled behind the wheel.

We got lost while trying to get out of the city, and then once we were finally on the interstate, Bill's voice came over the CB, saying, "Shit, this thing will need gas within a couple of hours. I should have filled up while you were back there playing Goldilocks."

I chastised him. "Shut up, dwarf. You just be grateful that your job is such a goddamn fairy tale. Not everyone can get away with biting the snow-white hand that feeds them the way that you do."

"Yeah, well, these trucks don't load themselves. So *you* just remember who your real friends are," he responded.

I smiled, noticing the slogan on the Pennsylvania license plate of the U-Haul that Bill was driving ("America Starts Here!"), but did not answer. I slipped a disc into the car stereo: *Songs from "Dawson's Creek."* I held the "talk" button down on the CB's microphone and wrapped electrical tape around it to keep it engaged. Then I set it down carefully, directly in front of one of my car's audio speakers, secure in the expectation that Bill would be driven totally batshit crazy by the third track of those bubblegum pop songs. We got in the slow lane and drove eastward, unsure of who was following whom.

3

FOR TREES THAT LIVE in the snow, winter is a journey. Plants do not travel through space as we do: as a rule they do not move from place to place. Instead they travel through time, enduring one event after the other, and in this sense, winter is a particularly long trip. Trees follow the standard advice given for any extended travel within a rustic setting: pack carefully.

Remaining stationary and naked outside in the below-freezing weather for three months is a death sentence for almost every living thing on Earth, except for the many species of trees that have been doing it for a hundred million years or more. Spruce, pine, birch, and the other species that blanket Alaska, Canada, Scandinavia, and Russia endure up to six months of frozen weather each year.

It may not surprise you to learn that the whole trick of survival is not freezing to death. Living organisms are made mostly of water, and trees are no exception. Every cell within a tree is basically just a box of water, and water freezes at exactly zero degrees Celsius. Water also expands as it freezes—the exact opposite of what most liquids do—and this expansion can burst whatever the water is contained within. You see this if the back of your refrigerator is a bit too cold: after a slight frost, your celery is reduced to a limp, watery mess. This is because the cell walls have burst as the cell water froze, and this has ruined your vegetable.

Animal cells can tolerate frozen temperatures for short periods of time because they are constantly burning sugar to produce energy in the form of heat. Plants, in contrast, make sugar, taking in energy in the form of light. If the sun is not strong enough to keep the air above freezing, then the tree is not kept above freezing either. The Earth's rotation is such that the North Pole tilts away from the sun for part of each year, reducing the amount of heat that is supplied to the high latitudes, and this is what causes winter in the Northern Hemisphere.

In order to prepare for their long winter journey, trees undergo a process known as "hardening." First the permeability of the cell walls increases drastically, allowing pure water to flow out while concentrating the sugars, proteins, and acids left behind. These chemicals act as a potent antifreeze, such that the cell can now dip well below freezing and the fluid inside of it will still persist in a syrupy liquid form. The spaces between the cells are now filled with an ultrapure distillate of cell water, so pure that there are no stray atoms upon which an ice crystal could nucleate and grow. Ice is a three-dimensional crystal of molecules, and freezing requires a nucleation spot—some chemical aberration upon which the pattern may start to build. Pure water devoid of any such site may be "super-cooled" to forty degrees below zero and still remain an ice-free liquid. It is in this "hardened" state, with some cells packed full of chemicals and others sectioned off for purity, that a tree embarks on its winter journey, standing unmoved through the frost, sleet, and blizzards of the season. These trees do not grow during winter; they merely stand and ride planet Earth to the other side of the sun, where the North Pole will finally be tilted toward the heat source and the tree will experience summer.

The vast majority of northern trees prepare well for their wintertime journey, and death due to frost damage is extremely rare. A chilly autumn brings on the same hardening as a balmy one, because the trees do not take their cue from the changing temperature. It is the gradual shortening of the days, sensed as a steady decrease in light during each twenty-four-hour cycle, that triggers hardening. Unlike the overall character of winter, which may be mild one year and punishing the next, the pattern of how light changes through the autumn is exactly the same every year.

Multiple light experiments have shown that the changing "photoperiod" is what triggers the tree to harden; it can be triggered in July if we fool the tree using artificial light. Hardening has worked for eons because a tree can trust the sun to tell it when winter is coming, even during years when the weather is capricious. These plants know that when your world is changing rapidly, it is important to have identified the one thing that you can always count upon.

4

I WAS COVERED in dried, flattened leaves. My hair was full of them, and I could feel crunchy bits of stem on my scalp, slipping down into my collar. There were shreds of leaves stuffed into my boots and they found their way to the insides of my socks. My wrists were stained black with the grime of the dry leaf dust that rubbed itself in whenever I pulled my gloves off or on. Sneezing yielded a smear of mucus with leaf mulch in it and I could taste the grit of dry, dead leaves in my mouth. Every time I reached up with my knife a gush of compressed dried leaves rained down on me. I didn't even bother trying to keep the debris out of my eyes; I just shut them tightly while I dug.

Bill and I were spending the summer about seven hundred miles north of the northern coast of Alaska, on Axel Heiberg Island, which is part of the vast Nunavut territory of Canada. Thanks to our GPS we knew exactly where we were on the globe, within inches actually, and yet our overwhelming feeling was that we were completely off the map. Our group of twelve scientists represented the only human beings within a three-hundred-mile radius. The Canadian military flew in every few weeks to check up on us, but in between their visits we were completely alone with our thoughts and with one another.

One of the strangest things about being thousands of miles from anywhere is how incredibly safe you feel. Nothing surprising is going to happen. You aren't going to bump into anyone you don't know. Water bleeds from the melting permafrost, making the earth spongy and so soft that you can't even fall down and hurt yourself. In theory, hungry polar bears could wander inland and eat you, but the scientists I know who have been working there for more than a decade tell me that they have yet to see one that far inland.

The landscape is flat and you can see for ten miles or more because the air is crystal clear. There's no grass, no bushes, and certainly no trees. You don't see many animals either, because there's almost nothing for them to eat. The life forms that you do come across—a lichen glued to a rock, a single musk ox trudging across the landscape, a nondescript bird passing far overhead—are few and far between.

The sun never, ever goes down. It just endlessly circles you, low in the sky, as if it were riding a merry-go-round with you standing at the center. Life is quiet and surreal. You abandon your habit of keeping track of what day and what time it is. You sleep until you wake up, you eat until you're full, and you work until you're tired—trading back and forth among these three activities. No matter how long you work in the Arctic during the summer, you are there for exactly one day. Then you go home in order to avoid winter: a three-month night during which the sun never rises. You won't be there, but that lichen, that bird, and that musk ox will be, stumbling around in the dark and still searching for something to eat.

The place where we work in the Arctic is more than one thousand miles away from the nearest tree, but it wasn't always like that. Canada and Siberia are loaded with the remains of what were lush deciduous conifer forests that sprawled north of the Arctic Circle for tens of millions of years, starting about fifty million years ago. Tree-dwelling rodents climbed the branches of these forests and looked down upon huge tortoises and alligator-like reptiles. All of these animals are now extinct, but together they formed an ecosystem more reminiscent of Alice's Wonderland than of anything that can be found today. It's obvious that the climate of the polar regions was warmer during that time, and that there were certainly no frozen fields of unforgiving ice, like there are today.

What puzzles us as botanists, however, is that those forests somehow persisted through three months of total darkness each winter, to be followed by three months of continuous summer sun. Extreme light regimes are incredibly stressful to today's plants, which generally would not live through a year of such treatment. In contrast, forty-five million years ago, the Arctic was home to thousands of miles of dense, productive deciduous forests that thrived through these wild swings in illumination. The discovery of trees that could live in the dark is akin to a discovery of humans that could live underwater. We must conclude that either the trees of the past were capable of something that today's trees cannot do, or the trees of today no longer use these talents and instead keep them tucked away as an adaptive ace up their evolutionary sleeve.

Bill and I, and ten other researchers who hailed from the paleontology department of the University of Pennsylvania, had been set down on Axel Heiberg in groups of four, delivered by helicopter, after riding a twin-engine aircraft, after riding plane after plane, moving north from Toronto to Yellowknife to Resolute and on up for days. Standing in the mud, watching the helicopter fly off, we looked down at our backpacks and up at each other, and realized how profoundly alone we were in our little group.

Over the next five weeks, the paleontologists would spend day after day parked in one spot, carefully exhuming single specimens of the buried fossil trees. They worked painstakingly, basically digging a trench with ten toothbrushes. They uncovered amazing fossils: tree trunks six feet in diameter and almost perfectly intact. Because the ground was frozen, the sediment covering the fossils had to be scraped off centimeter by centimeter once the sun had thawed the top layer; it was like digging through ice cream that had frozen too hard for easy serving. The paleontologists had a few different specimens that they were unearthing, and they used small cards of plastic to do it, similar to the way that you might scrape ice off of a windshield using your driver's license. They rotated among fossils, the perpetual sun slowly assisting them.

The fossils were still made of wood, which is what made them precious. Most of the tree fossils you can think of have been petrified as fluids have passed through them for ages, swapping molecule for mineral until the tree has fully turned to rock. In contrast, the fossils of Axel Heiberg Island contained wood tissue that was still intact—you could even burn the fossils to heat your bathwater, which is what the rough-and-ready grizzly male geologists had done with it during the eighties when the site was first surveyed, if legends are to be believed.

The paleontologists of our trip were a more housebroken version of the classic mountain-man geologist, but still hard-working, hard-drinking, and fascinated with the gun that the Canadian government required us to carry in case of polar bears. I had learned to keep my distance from these types of colleagues, knowing that they would never accept me as having a legitimate intellectual claim to the site, even if our funding agency did. In their eyes, I was just a grubby little girl who couldn't lift forty pounds with a weirdo in tow, and I embraced this role, hoping it would result in their underestimating me to the point of leaving me alone. Somehow, all of our sleep cycles settled into a pattern where Bill and I worked while they slept and vice versa.

Bill and I also took a fundamentally different approach to the site, as compared with our more established colleagues. I was obsessed not with the fantastic individual fossils but with the incredible duration and stability of the forest as a whole. This was no flash-in-the-pan freaky ecosystem; this configuration of global biology had persisted for many millions of years—millions of years during which huge amounts of carbon and water flooded into the Arctic and were transformed into leaves and woods, and then shed annually in a gush of tissue. How in the world did the system sustain itself? There is nowhere near that kind of fresh, liquid water available in the Arctic now, not to mention the lack of soil nutrients.

Bill and I decided that instead of looking at a single snapshot of time by digging up a few individual trunks, we were going to tunnel vertically through the whole mess, looking for subtle changes through time in the chemistry of the mummified wood, leaves, and sticks. This meant digging apart and sampling layer after layer of the dead, compressed debris that had accumulated over millions of years. As we burrowed vertically through cross sections of dry, rotted leaves, we sampled centimeter by centimeter and recorded precisely where in the column we were. By the end of three summer field seasons, we'd sampled our way through a hundred vertical feet of time and were able to identify at least one strong swing in climate that the forests had been able to tolerate. From this we have argued that these ancient Arctic ecosystems are better characterized as "resilient" than "stable."

We chose a spot across the basin, far from the place where the paleontologists were excavating, and for weeks on end, Bill and I dug up through sediment layers that were more than ten feet thick and interbedded with gravel and silt. Each week we'd thrash around in a different twelve-foot pile of forty-million-year-old dry compost. Often we worked while hanging off of one side of a gentle, sloping, and soft cliff that was continuously giving way out from under us, sending us tumbling down the hill covered in a debris slide.

We dug without secure footing while we tried to get a clean sample and keep track of our position relative to a base elevation. But in this environment it was difficult to the point of being silly, and we cycled back and forth between riotous laughter and rageful frustration throughout those long days of cartwheeling down the hills. One time when I dug up with the claw end of my hammer I cracked something odd and pounds of sparkling clear amber rained down on my head. "So this is what it's like to be an earthworm," Bill remarked once after a particularly big avalanche, and I remember pausing in appreciation of how his observations were always so spot-on.

At least once a day, we indulged ourselves in the following way: we'd plop down waist deep in the crunchy rubble and pull out some treats. Nothing tastes as good as a Snickers bar and a hot thermos of coffee in the cold middle of nowhere, and once a day we focused all of our energy toward savoring this pleasure in quiet, companionable reflection.

One day, after we had chewed our last bites, Bill raised his arm and silently pointed at a gray speck many yards off. I puzzled for a moment and then saw that he was pointing out an Arctic hare. Coming across an animal—any animal—is a rare treat in the Arctic, because an herbivore must travel long distances to keep itself fed upon the sparse moss and lichen available, and a carnivore must, by extension, keep moving to follow wide-roaming prey.

The hare came closer, picking among the rocks, and then began to move away from us. Bill and I stood up and followed it, keeping a good distance and leaving our gear behind. We walked for about a mile without talking, following the hare and watching it, trying to exploit the visual novelty that it offered against the bleak and monotonous landscape. It was a large hare, big as a sheltie dog and with similar fur, long ears, and a long, lean body. It didn't seem to mind us following it at a distance of a quarter mile or so, so we hung back and followed for more than an hour. There was no real way for us to get lost; we could walk all day and still turn around and see the Day-Glo orange tents of our campsite.

When you are supremely isolated among just a handful of people, those few people can quickly begin to seem suffocating. And they did—except for Bill, I had discovered. Before that trip, I had never really spent twenty-four-seven with anyone for weeks on end, and it seemed that with every passing day it got easier for us and not harder. Regardless of whether we were awake or asleep, we were never more than a few feet apart, though always in separate tents. Some days we chattered incessantly; some days we said only a few words, and then we lost track of what we did or didn't say, of how much we were talking or not talking. We were just us being us.

On the day that we followed the hare, we eventually found ourselves on a geographical high point, and I turned to see that our colleagues back at the dig were only fuzzy specks in the distance, as we must have been to them. In the other direction, we could see the edge of a glacier that lay like a thick white layer of frosting, still several miles away. I sat down to admire it and Bill sat down a few feet away from me. We sat in silence for another half hour, until Bill finally said, "It feels weird not to be working."

"I know what you mean," I said. "And we've dug through every layer twice while sampling. It doesn't make any sense to do it again."

"But we have to do something," countered Bill. "Otherwise the Grizzly Adamses down there are going to wonder what the hell we're doing on this trip, won't they?"

I laughed. "They already wonder what I'm doing here. And digging to China and back wouldn't convince them that I'm a legitimate scientist."

"Really?" Bill looked at me in surprise. "I always figured it was just me who felt like a random mistake."

"Naw," I assured him. "Look at those guys. I'm going to do this job for thirty more years, work as hard as any of them, accomplish just as much or more, and not one of them will ever look me straight in the eye like I belong here."

"Well, at least you have two whole hands," Bill countered, wiggling his incomplete set of digits. "That's a good start anyway."

I lay back and looked at the sky. "Oh, come on, nobody ever notices your hand," I said. "Honestly, you're more normal-looking than anyone I know; I don't know why you don't get that."

"Are you sure? Why don't you poll some little kids about it?" Bill asked. "Like my second-grade class. And third grade. And high school, and so on."

I sat up with a start. "They teased you? At school? About your hand?" The idea enraged me.

"Yup," confirmed Bill quietly, still staring at the sky.

I pushed the point. "So is that your deal? You've carried that all these years? Is that it? Living in a hole, no friends?"

"That's about right," Bill confirmed.

"You never did Cub Scouts, joined a team, all that crap?" I listed all the usual milestones that I had taken for granted.

"Now you're getting it," Bill acknowledged.

"You've never been on a date, have you?" I asked. It seemed to have been hanging in the air, and it just felt right to say it.

Bill stood up and raised both arms into the endless blue-white sky that seemed, on that bright July day, to be incapable of darkness. *"I never went to the prom!"* he howled.

When our laughter died down, I thought a little and then spoke. "So why not now?" I suggested. "We're in the middle of nowhere, with no one to see you. You could dance now."

There was a long pause. "I don't know how," said Bill.

"Yes, you do," I insisted. "It's not too late. C'mon, we came all this way. Jesus, *this* is why we're here. I just figured it out. This is the place where you dance."

To my surprise, Bill didn't joke it off this time. He took a few steps toward the glacier; he stood and stared at it for a long time with his back to me. Then he slowly began to turn in a circle and stomp, and do rough jumping skips in between stomps. It started out awkward, but he soon threw himself into it, spinning and stomping and jumping. Soon he was moving with abandon, but deliberately, not frantically.

I sat directly in front of him, held my head up, and watched. I watched him, as a clear-eyed witness of what he was doing and of what he was, of all of it. There at the end of the world, he danced in the broad and endless daylight, and I accepted him for what he was, instead of for what he wished he could be. The potency of my acceptance made me wonder, just a little, if I could turn it inward and accept myself. I didn't know, but I promised myself that I would figure it out on another day. Today was already taken. Today was for watching a great man dance in the snow.

5

ALL OF THE SEX on planet Earth is biologically designed to serve one evolutionary purpose: to mix the genes of two separate individuals and then produce a new individual sporting genes identical to neither parent. Within this new mix of genes are unprecedented possibilities, old weaknesses eliminated, and new weaknesses that might even turn out to be strengths. This is the mechanism by which the wheels of evolution turn.

All sex involves touch: the living tissues of two separate individuals must come into contact and then attach. Contacting and attaching to another individual is a big problem for plants: they are anchored in place and their survival depends upon their immobility. Nevertheless, the vast majority of plants faithfully produce a new crop of flowers every single year, fulfilling their half of the reproductive bargain, even though the odds of these flowers eventually becoming fertilized are small.

Most flowers are built simply: a platform of petals surrounding the "male" and "female" parts. On the outer ring of this circle is the male component: a few long stalks with wads of pollen glued loosely to the ends. In the center and at the bottom of a chute sit the ovaries. Of all the things that could travel down this open chute, a grain of pollen from the same plant species is the only thing that might activate fertilization. Self-fertilization is slightly more likely to occur, which means that the ovules will come in contact with pollen from the same flower. This can result in a seed, and then possibly a new individual, but no new genes will have been introduced. For the species to persist and evolve, real fertilization must happen periodically, and this means that pollen from one, or ten, or ten thousand feet away must successfully arrive at the ovaries.

There is a wasp that cannot reproduce outside the flower of a fig; this same fig flower cannot be fertilized without the help of a wasp. When the female wasp lays her eggs inside the fig flower, she also deposits the pollen that coated her when she hatched within a different fig flower. These two organisms—the wasp and the fig—have enjoyed this arrangement for almost ninety million years, evolving together through the extinction of the dinosaurs and across multiple ice ages. Theirs is like any epic love story, in that part of the appeal lies in its impossibility.

Such specificity is extremely rare in the plant world, so rare as to be hardly worth mentioning, except as a feel-good example of symbiosis between ecological soul mates. Much more than 99.9 percent of pollen produced in the world goes absolutely nowhere and fertilizes nothing. As for the infinitesimal number of grains that arrive in the right place, it seems irrelevant to claim that it matters how they got there. Wind, insects, birds, rodents, or the empty corners of FedEx boxes—the vast majority of plants have absolutely no preference as to the method of pollen delivery.

Magnolia, maple, dogwood, willow, cherry, and apple trees—they spread their pollen on every kind of fly or beetle, luring them near with sweet nectar but providing only enough for a brief taste. The value of an insect as a pollinator lies in the distance that it may travel, and thus less time spent lounging about on a petal means more time traveling in the air. Many shrubs in North America and Europe bear flowers with petals set to spring once weighted by an insect, smacking the bee full of pollen and back on his way.

In contrast, the elm, birch, oak, poplar, walnut, pine, and spruce, as well as the grasses, all release their pollen to the wind. It goes farther than it would by insect, but never as directly to another flower. Windborne pollen travels for miles and then rains down indiscriminately. Enough of it hits its target, however, to keep the world perpetually blanketed by the great conifer forests of Canada, the Giant Redwood groves of the Pacific Northwest, and the expansive spruce forests that stretch through Scandinavia and Siberia.

One grain of pollen is all that is necessary to fertilize an ovum and then develop into a seed. One seed may grow into a tree. One tree can produce one hundred thousand flowers each year. Each flower can produce one hundred thousand grains of pollen. Successful plant sex may be rare, but when it does happen it triggers a supernova of new possibilities.

6

WHEN I WAS THIRTY-TWO I learned that life can change in one day.

Within certain social circles of the married, a single woman over the age of thirty inspires compassion similar to that bestowed upon a big, friendly stray dog. Although the dog's unkempt appearance and tendency toward self-reliance betray its lack of an owner, the way it gravitates hungrily toward human contact suggests that it might once have known better days. You consider letting it eat on the porch after you confirm that it is not mangy, but then you decide not to, vaguely worried that it might start hanging around because it has nowhere else to go.

In the right venue—at a casual outdoor picnic, perhaps—a stray dog is a curiosity and even an asset; its muddy clowning provides a rosy window into the carefree life of a simpler being. As everybody's pet and nobody's responsibility, it is at least friendly, if not wholesome, and is remarkably happy given its humble lot. If a single woman can be thought of as a dog at such events, then a thirtysomething single man is effectively characterized as the guy manning the hamburger grill. He's sure to be plagued by the dog from start to finish, whether he likes animals or not.

I met Clint at a barbeque of that sort, and he couldn't have shooed me away even if he had tried, because he was easily the most beautiful man I had ever seen. A week later, I worked up the nerve to ask our hostess for his e-mail address, and then I wrote and invited him to dinner. After he accepted, I phoned to tell him the location, which was the trendiest restaurant I could identify near Dupont Circle. I'd certainly never been there, but it seemed like a place where people went on fancy dates, and Washington, D.C., was much cooler than Baltimore—I did know that. After giving him directions I further stipulated, "I'll only show up if you agree that I am paying for dinner." I had always paid my own way through life, and I wasn't about to give that up now.

"Okay." He laughed good-naturedly. "But you have to let me pay the next time." I didn't promise anything, but I took his words as a good omen.

At dinner I couldn't eat a thing because I didn't want to be distracted from the fact that something wonderful was happening. We left the restaurant laughing over our waiters' disapproving stares during the course of our three-hour meal. We went to a pub a few blocks away and talked for hours while our drinks went untouched. We argued over the essential difference between measuring something and modeling it. We discussed mosses and ferns. It turned out that we had attended Berkeley and studied the same subjects at the same time. I knew many of his friends and classmates, and he happened to know many of mine. We even determined that we had sat in the same room on more than one occasion, both listening to the same seminar. We marveled over how in the world we could have missed each other all these years. It seemed obvious that we should make up for it now.

They closed the bar and I still didn't want to go home. We decided to go to his house, and he asked me if I wanted to walk or take a taxi. He saw the look on my face and then stepped into the street to wave for a taxi. Where I grew up, we only ever saw taxis in the movies. Taxis were for people so sophisticated that they left their houses wearing shoes they couldn't even walk in. Taxi drivers were exotic guides into the unknown who dispensed detached wisdom while delivering you faithfully to an important place that you couldn't have found on your own. I was stunned to find that the ultimate proof of love for me was nothing heroic, but an easy and superfluous gesture performed just to make me smile. The love that I had to give someone had been packed too tightly and too long in a small box, and so it all tumbled out when opened. And there was more where that came from.

We love each other because we can't help it. We don't work at it and we don't sacrifice for it. It is easy and all the sweeter to me because it is so undeserved. I discover within a second context that when something just won't work, moving heaven and earth often won't make it work—and similarly, there are some things that you just can't screw up. I know that I could live without him: I have my own work, my own mission, and my own money. But I don't want to. I *really* don't want to. We make plans: he will share his strength with me and I will share my imagination with him, and in each other we will find a dear use for our respectively obscene surpluses. We will fly off to Copenhagen for the weekend and live in the south of France each summer; we will get married in a language that we don't understand; I will have a horse (a brown mare named "Sugar"); we will go to avant-garde theater productions and discuss them afterward with strangers in coffeehouses; I will give birth to twins like my grandmother, but we'll keep the dog (duh) and we will always take taxis and live like in the movies. And some of these things we do, and some of them we don't (like the horse), and it is better than a movie, because it doesn't end, and we are not acting, and I am not wearing any makeup.

* * *

Within a couple of weeks I had convinced Clint to quit his job in D.C. and move into my house in Baltimore, knowing that his incredible talent for mathematics could get him a job anywhere. Soon after moving, he reentered academia and took a job at Johns Hopkins researching the deep Earth, in the same building as my lab. He spent his days writing fantastically intricate computer models designed to predict million-year flow within unfathomably hot and pressurized pseudo-solid rock, thousands of miles below the depth where the lava of a volcano brews. I couldn't understand—still can't—how he was able to study the Earth in his mind only, how he could imagine and observe its workings through the baroque equations that he writes so fluently, the corner of his mouth always inky from the ballpoint pen that he doesn't notice he's been chewing.

I have to see my science for it to be real: I must hold it in my hands and manipulate it; I need to watch plants grow and make them die. I need the answers that can come only from control; he prefers to set the world in motion and then watch it flow. Tall and thin and dressed in khaki, he looks and acts just as a scientist is expected to, and so acceptance into the profession has always been relatively easy for him. Nevertheless, his sweet, solid, and loving nature was a treasure overlooked until I recognized it and then decided to never, ever let it go.

Clint and I met in early 2001, and during the summer that followed, we took a trip to Norway, so that I could show him the places that I love most: long, low hills of pink granite with purple wildflowers pushing out of the cracks, sparkling fjords superintended by sober-faced puffins, white birch trees illuminated by salmon-colored sunsets that last all night long. The Oslo leg of our trip was transformed into an impromptu wedding party after we took a number, waited in a queue for twenty minutes, and got married in the Rådhus (City Hall).

Upon returning home to Baltimore, we went straight to Bill to surprise him with the good news. Bill had never commented on the guys that I dated, probably because it was obvious that there wouldn't be very many dates to comment on. But he'd been acting strangely since Clint had come on the scene, avoiding us the same way that a reformed felon avoids driving by the hoosegow. For his part, Clint was confident that Bill just needed time to get used to the situation; it was exactly the same as with me and his three little sisters, he kept insisting.

About a month earlier, Bill had moved out of my attic after buying the run-down house located just a few doors down. He now owned a four-floor row home that must have been beautiful in its day, but that day was long past. When he moved in, Bill had dumped all of his belongings on the first floor, having carried them over from my house one by one over the course of several days. He kept a few key items (coffeepot, razor, screwdriver) in a corner next to the nest of laundry that he crawled into and back out of when it was time to sleep and wake up. Bill had grand renovation plans for the place, but during that summer it looked like a heroin den, complete in every detail except for the drugs.

The day after we returned from Norway, we knocked hard on Bill's door and then also rang the doorbell. At length we heard someone shuffling around, and a bit later we heard the lock turn. The door opened and there stood Bill in a ripped T-shirt and some faded swimming trunks. His hair was mussed and he was rubbing his eyes: we'd obviously woken him up. It was three o'clock in the afternoon.

"Hi!" I greeted him, standing there with Clint's arm around me. Then I burst out, "Guess what? We got married!"

There was a long pause while Bill looked at us blankly. "Does this mean I have to buy you a present?" he asked.

"No," answered Clint, while I simultaneously responded, "Yes."

We stood there for a while, Clint and I with giddy smiles on our faces. Finally I said to Bill, "Get dressed. There's a Civil War reenactment downtown at Fort McHenry and we're going."

"I'd join you, except that it's probably the War of 1812, you're a dumbass, and I have about a million other things to do," Bill answered, looking uncomfortable.

"Watch your mouth, you filthy hippy," I chastised him. "I won't have you dishonoring our fallen heroes that way." I added, "Now, put some fucking pants on, start acting like an American, and get into the Toyota."

Bill still looked at us, and I knew he was conflicted over whether to consent or withdraw. I looked up at my new husband, the strongest and kindest man I had ever met, firm in my belief that everyone who had ever earned my love had a natural claim to his as well.

"C'mon, Bill, you're with us now," Clint said while offering his car keys. "Why don't you drive?" he added. Bill took the keys and we spent the day at Fort McHenry bobbing for apples and dipping candles and forging a real horseshoe. We ate hot dogs and cotton candy and watched the three-legged race and petted animals at the petting zoo. And we all got in for a reduced rate because it was, after all, Family Day.

7

AGRONOMISTS AND FORESTERS have charted the growth of hundreds of plant species, starting in 1879 when a German scientist noticed that the increasing weight of a corn plant, when graphed against the days of its development, resulted in a line with a curious lazy-S shape. These scientists had weighed their potted plants daily, and for the entire first month they saw very little growth. Then, during the second month, the plants' weights shot up sharply; they doubled in size each week until they reached their maxima at three months of age. The scientists were then surprised to see the weights drop off again, and by the time they began to flower and produce seed, the plants weighed only about 80 percent of what they had been at their largest. This scientific result was enduring, and the many thousand corn plants that have been charted since then have all shown a similar lazy-S curve. We don't know exactly how it works, but a corn plant knows what it is supposed to be, even though it meanders along the way.

Other plants display very different growth curves. The curve describing the development of the leaves on wheat resembles your pulse: a brief throb of growth that settles back into decline. The curve for a sugar beet also shows increase followed by decline; however, the curve makes a long, low arc centered on the summer solstice. The curve for the nuisance grass *Phragmites* looks like a pyramid: birth and growth symmetrically bracketed by decay and death. Such curves are invaluable in the farm field or in the forest, where harvesting food or wood is a distant goal. By approximating the position of growing plants along these standard growth curves, one can guess a good date for harvest, and by extension, pencil in the possible payday associated with it.

The growth curves of trees are diffuse and sprawling compared with the curves for smaller plants, as they extend for hundreds of years instead of for just one season. Each tree species is subject to its own unique curve. Monterey pine grows twice as fast as Norway spruce, but both trees are harvested for papermaking when they are of similar girth. As a result, Norwegian paper companies are more likely to be solvent and generally own larger territories than their American counterparts.

Within the forest, the variation in stature among trees of the same age is far larger than it is for other organisms, including animals. Within the United States, the tallest ten-year-old boy is about 20 percent taller than the shortest ten-year-old boy. This same differential holds for five-year-old boys as well as for twenty-year-old men: the tallest is about 20 percent taller than the shortest. Within a pine forest, the thickest ten-year-old trunk is about four times thicker than the thinnest ten-year-old trunk. This same differential holds for twenty-year-old trees as well as for fifty-year-old trees: the thickest trunk is about four times thicker than the thinnest. It turns out that there is no "right" or "wrong" way to grow into a hundred-year-old tree: there are only ways that work and ways that do not.

Becoming a tree is a long journey, and so even the most experienced botanist cannot look at a twig on a sapling and accurately describe what kind of branch it might develop into over the next fifty years. Plant growth curves can be useful for guessing, but it is important to remember that they don't show us the future, only the past. They are improvised lines, drawn through data that was collected for plants that are mostly dead by now. The datasets that define these curves are not static, and every time a new plant is measured it can be added to the graph. Each new data point changes the overall pattern slightly and thus alters the growth curve. There is no way to mathematically predict the shape of these curves, even with the massive computers that can lately be brought to bear. Nothing in these growth curves tells us what a tree should look like, only what trees have looked like. Every plant must find its own unique path to maturity.

There are botany textbooks that contain pages and pages of growth curves, but it is always the lazy-S-shaped ones that confuse my students the most. Why would a plant decrease in mass just when it is nearing its plateau of maximum productivity? I remind them that this shrinking has proved to be a signal of reproduction. As the green plants reach maturity, some of their nutrients are pulled back and repurposed toward flowers and seeds. Production of the new generation comes at a significant cost to the parent, and you can see it in a cornfield, even from a great distance.

8

BEING PREGNANT IS by far the hardest thing that I have ever done. I can't breathe, I can't sit down and I can't stand up, I can't put down the tray-table on the airplane, I can't sleep on my stomach, and I've *only* slept on my stomach for the last thirty-four years. I wonder what kind of god in what kind of heaven decided that a hundred-and-ten-pound woman could carry thirty-five pounds of baby. I am compelled to march in endless circles around the neighborhood escorted by Reba, because the baby is quiet only when I am moving. He kicks me not with here-I-am-Mommy playful bops, but in the writhing torture of a man struggling against a straitjacket. I walk and walk and walk some more, a solitary parody of some pagan fertility parade, and I think about how neither I nor the baby is enjoying this suffocating arrangement.

A manic-depressive pregnant woman cannot take Depakote or Tegretol or Seroquel or lithium or Risperdal or any of the other things that she's been taking on a daily basis for years in order to keep herself from hearing voices and banging her head against the wall. Once her pregnancy is confirmed she must cease all medications quickly (another known trigger) and stand on the train tracks just waiting for the locomotive to hit. The statistics are pretty simple: a bipolar woman is seven times as likely to experience a major episode while pregnant, compared with before or after. Leaving her to ride it out without medication for the first two trimesters is the cruel reality upon which doctors insist.

Early in the pregnancy I wake up and vomit violently until I collapse onto the bathroom floor and lie there for hours, retching and crying in exhaustion until finally, in desperation, I begin to hit my head against the walls and the floor, trying to knock myself out. I regress to my child's habit of begging Jesus for help or at least for the mercy of oblivion. Later, when I come to, I can feel a cool film of snot, blood, saliva, and tears between my face and the tile floor, but I cannot speak and I do not know who I am. My steadfast husband, who has been frantically on the phone, comes in and picks me up and washes me off, and calls the doctor again. They take me in and try all the things that they have tried on me before, but in a week I am right back to doing it again. This goes on until Clint and the dog are the only beings in the whole world whom I can recognize by name.

I go to the hospital in earnest and stay for weeks at a time, strapped down when nothing else works, and they put me through countless rounds of electroconvulsive therapy, which make me forget most of 2002. I beg the doctors and nurses to tell me why, why, why this is happening to me, and they do not answer. All of us can do little except count the days until it will be safe for me to take the drugs that I need. Twenty-six weeks is a magic date: it ushers in the third trimester, a period of advanced fetal development for which the Food and Drug Administration has approved the use of a whole series of antipsychotic drugs to address the health of the mother.

As soon as it is medically advisable, I am put on this and that and the other medication regimen, and then slowly my more florid symptoms begin to come under control. I begin to drag myself into work, often just to spend the day sleeping on the floor of my office. I try but find myself too weak to teach, and so I put myself on medical leave. One morning during my eighth month, I trudge through the front door of the building and stop to rest in the front office while mentally preparing to drag this extra thirty pounds down to my laboratory in the basement. I don't handle any chemicals, of course, but it comforts me to sit next to the humming machines and examine the readouts as they are produced, and pretend that the instruments need my approval and encouragement in order to continue with each next task.

In preparation for my difficult journey down in the elevator, I sit on one of the visitor's chairs next to the photocopier and lean back behind my huge abdomen. I announce, "I think I get it now. This is the new me. He is never coming out. Eighteen years from now I will have a grown man living inside my body," and although I don't really mean it as a joke, the secretaries chuckle in sympathy.

Walter, the head of the department, walks in, and I automatically stand up, like a soldier coming to attention in the presence of a senior officer. I am close to being the first and only woman ever awarded tenure in this hundred-year-old ivy-draped department at Hopkins, and I instinctively know that I should hide any physical weakness that accompanies my pregnant state.

Unfortunately, I've stood up too quickly and the blood rushes from my head and I feel faint. I automatically sit down and put my head between my legs, knowing that this will pass in a minute or so. This light-headedness is familiar because I have always been plagued by low blood pressure and tend to be resistant toward eating, viewing it as an endlessly recurring chore. Walter looks around in puzzlement, and at me in the middle of it all having assumed the posture of a beached and prostrate whale. He goes into his office and closes the door. Someone offers me a cup of water, but I refuse it. I limp to the elevator burdened by a new nagging worry that I can't quite put my finger on.

The next day at about six-thirty in the evening, Clint comes into my office, which is located just down the hall from his own. His face is drawn in such a way that I wonder if he has come to tell me that someone has died. He leans on the doorframe and says gravely, "Listen, Walter came into my office today." He pauses, looking pained. "He told me that you can't come into the building anymore while you are on medical leave."

"What?" I cry out, more terrified than angry. "How can they *do* that? It's *my* lab; I *built* that place—"

"I know, I know . . . ," sighs my husband. "They're assholes." But he says it gently, to soothe me too.

"I didn't know they could do *this,*" I respond as the hurt starts to sink in. "*Why?* Did he say *why?*" I ask, and the many, many times in my life when I have pleaded this question *"Why?"* to those in power come back to me, as does the fact that I have never, ever been given an answer that's good enough for me.

"Oh, some bullshit about liability and insurance," he answered, and then continued, "They're cavemen. We *knew* that."

I begin ranting, "What the fuck? Half of these guys are drunk in their offices . . . and hitting on students . . . and *I'm* the liability?"

"Listen, here's the reality. They don't want to look at a pregnant woman, and you're the only one who has ever set foot in this building. They can't deal. It's simple," he says softly, and his anger is calmer than mine.

Part of me is still flabbergasted. "He told *you* to tell *me*? Why didn't he come and tell *me* himself?"

"He's afraid of you, is my guess. They're all cowards."

I shake my head and clench my teeth. "No, no, *no*!" I insist.

"Hope, we can't *do* anything about this," he says quietly, wretchedly. "He's the *boss*." Clint is wearing the same soul-worried look that I once saw on a magnificent and ancient elephant that had lost its mate of thirty years. He knows how much it hurts me to be banned from my own lab, from the place where I feel happy and safe—especially now—and from the only place that I truly regard as home.

In frustration I grab my empty coffee cup and hurl it to the floor with all my might. It bounces on the carpet and does not break but instead rocks itself into a smug and leisurely sideways pose. In it I see yet more evidence of my powerlessness, even over things small and meaningless, and I sit down, put my head in my hands, and sob onto my desk.

"I don't want this anymore," I choke out while practically keening, and Clint stands and witnesses my pain, and the weight on his heart doubles, and doubles again. After my crying has abated, we sit together in silence and sufficient unto the day was the evil thereof.

Two years later Clint would tell me that all his affection for Hopkins died on that day, and that he had never forgiven them for hurting me. We talked about it with the benefit of distance and hindsight—how it probably was just for liability reasons and nobody's fault—but then we stood together, joined hands, gathered all of our loved ones and a few of our belongings, and moved thousands of miles away.

And yet again I built up my lab from zero, with Bill at the middle of it. But on the day that I throw my coffee cup, I cry because I can see only what I am losing and not what I will gain, which is hidden from me by my two-inch-thick uterus.

After I am banned from the department, I am at loose ends during the day, so I schedule my prenatal visits for the morning. I show up and the nurses and techs weigh me and ultrasound me and deliver the astounding news that I am one week more pregnant than I was one week ago. Strangers ask me how many "months along" I am, and when I answer "eleven" they expect me to laugh with them, but I fail even in this small thing.

I know that I am supposed to be happy and excited. I am supposed to be shopping and painting and talking lovingly to the baby inside me. I am supposed to celebrate the ripening fruit of love and luxuriate in the fullness of my womb. But I don't do any of this. Instead I grieve long and hard for the part of my life that is over now that this baby has come. I should revel in expectation, and daydream in circles about the mysterious identity of this person coming together inside me. But I don't, because I already know him. From the first, I sensed that he would be a boy, and I knew he would have blond hair and blue eyes like his father.

I understood that he would have my father's name and his own personality. That he would be as hardy as all Viking men and women are, and that he will justifiably hate me for being an unfit mother, that part of me having grown up under too much shade and wizened without flowering properly. I breathe in and out, I drink gallons of milk and eat buckets of spaghetti and sleep many hours each day, and I try to focus on the fact that I am at least sharing my rich blood with him, passively giving him what he needs, for now. I try not to think about my troubled mind. I try not to wonder when will be the next time that I lose my mind.

I sit in a waiting room with fifteen-year-old pregnant girls, each facing a wall of trouble much taller than mine, but I am numb to the gratitude that this should inspire. I am so sad that I cannot cry, and so empty that I cannot pray. The doctor calls me in and I notice that she is not wearing earrings. Neither am I. I reflect incongruously on the relative rarity with which I meet a woman who is not wearing earrings.

"Well, you're pretty big, but otherwise you're right where you should be," she announces, looking at my chart. "The baby's heartbeat is strong, and your blood sugar is normal. This is almost over," she says, and she looks at me hard. She hands me some pamphlets and asks, "Have you thought about contraception for after the birth? I suppose that you know that you can get pregnant even if you are breast-feeding."

My mind spins. This last stage of pregnancy has been positively surreal. Acquaintances ask me when I will have my second kid. Doctors prod me toward contraception. How bizarre to question a woman who can't even picture herself with one baby about the logistics (or not) of a second.

I stammer confusedly, "I don't think I can breast-feed. I mean, I have to work, and if I need medication or something—"

"It's okay," my doctor interrupts. "He will grow fine on formula. I am not worried about that."

Her forgiveness for the very first of my failures toward this baby is so automatic and freely given that it pierces me. I feel the old childish hope involuntarily stirred, that perhaps this woman cares and understands. After all, she has my chart. Maybe she noticed all the ECT and hospital stays and medications. Then I catch myself and listlessly wonder again for which of my sins I am being punished. I am sick to death of this wound that will not close; of how my babyish heart mistakes any simple kindness from a woman for a breadcrumb trail leading to the soft love of a mother or the fond approval of a grandmother. I am tired of carrying this dull orphan-pain, for though it has lost its power to surprise, every season it still reaps its harvest of hurt. *This woman is my doctor; she is not my mother,* I tell myself firmly, and I am humiliated in my need, even to myself. More immediate is the fact that someone somewhere who makes the schedules has already decreed that we have exactly twelve minutes for each other.

After we confirm my next appointment, I leave the doctor's office. On my way out, I go into the bathroom and I vomit and shake, and afterward I don't recognize the person in the mirror. She looks so sad and tired and greasy that I feel sorry for her, even before I fully understand that she is me.

After five o'clock, when everyone in the building has gone home for the day, I take Reba and sneak into the lab. I cannot do anything productive, but I instinctively resist the cruelty of my department chair's order by staging a sort of one-woman pregnant sit-in. When Bill walks in at seven-thirty, returning from his first meal of the day, to find me sitting in the dark, I hastily rub my face in order to hide the fact that I've been unable to stop crying. He turns on all the lights and systematically begins to update me on each of our projects, reciting a detailed status for each one, a long, comforting litany providing concrete evidence that everything is actually okay. He is exhausted from doing both of our jobs, but the rockier the ground, the harder he pulls the plow.

He doesn't know exactly what's been wrong with me or why I've been so absent. Neither do my friends or family, such as I have. And no one asks. I suppose that my lineage has been hiding the crazy for so many generations that my secrecy on these matters was genetically hardwired.

Bill assures me that it's okay for me to just stay home. "Seriously, no one is going to come in here; you don't have to guard the place at night." He looks around surreptitiously and adds, "Not with all the knives and shit I have in here," as he pretends to fumble nervously with one of the cabinets. This preposterous statement represents a new high-water mark in Bill's desperate efforts to make me laugh, or at least to provoke a sign of the old me when our paths do cross. Although we are both at a loss as to how to kill the despondent zombie that has taken over his best friend's ballooning frame, he does keep trying.

"Jesus, you look miserable," he says. "Why don't you go slaughter a pig or something? Isn't that what makes your people happy?" Bill is exasperated.

"Well, I am hungry . . . ," I offer.

With much effort, we walk (I waddle) to Bill's house and then watch reruns of *The Sopranos* while I eat the box of doughnuts that we bought on the way. At nine o'clock, when Clint comes to pick me up and drive me the three blocks home, he opens the back door and hands me in, and tears roll down my face as we pretend that our car is a taxi.

It is a good omen when you observe an experiment closely, prepared for the data to be subtle, but what you see is clear, forceful, and obvious beyond misinterpretation. I have been warned repeatedly that water-breaking could be ambiguous, but later that evening while sitting on the couch, I spontaneously find myself immersed in about a gallon of fluid. As the tide rises, I set my jaw and remark to Clint that we should probably go to the hospital.

As he helps me up, he notices that my hands are shaking. "We are going to the best hospital in the world," he reminds me calmly, and his confidence proves contagious. I gather my weak determination and we pack up and drive downtown. It is about ten-thirty in the evening, and as we pass through the miles of Baltimore urban housing projects I see people trudging home after their long day out, needing rest but not anticipating any.

We enter the hospital, and I am immediately comforted by the bright lights and buzzing activity, and strangely enough, an old feeling of safety comes back from my hospital pharmacy days. Each one of these busy people has a mission, and taking care of me is just a routine part of their huge, choreographed collective task. Whatever happens, I will not be alone, and somebody else will be the strong, prepared, alert, and responsible one. A plan is coming together: everybody will stay up all night and we will figure this thing out. I start to relax.

We share the elevator up to the maternity ward with an older patient who is being wheeled elsewhere by a young and bored orderly. She looks at my mammoth abdomen. "You ready for this?" she asks, and then shakes her head in wry amusement when I stare back at her dumbly, unable to formulate an answer.

When we get to the registration desk, a huge woman swoops in after eyeing me and says to the receptionist, "I want her; she's got good veins," and thus designates herself as my nurse. I look at the backs of my hands that are like my father's and upon which the blood vessels have always stood out clearly, and I decide that this is also a good omen. The nurse leads us to a private room and ushers Clint toward a chair in the corner; he is to sit near the foot of the bed and keep himself out of the way. He complies.

"This is not about *you,*" she explains to him over her shoulder as she guides me into the bathroom.

With great effort, I use the toilet and then change into a hospital gown. The nurse helps me up onto the bed and swabs both of my wrists with alcohol. She then pulls out ten or twenty needles, electrodes, clamps, and bands and begins to attach them to me in myriad ways and places. After she finishes she plugs each one separately into the machines and monitors that have begun to crowd around my bed as if they are eager to be included in whatever is going on. Once everything is turned on, I am surrounded by their friendly electronic faces on all sides, and each continuously repeats to me its own soothing story, as if they all understand that there is no upper limit to the amount of reassurance that I will require during my ordeal.

The physician's assistant walks in. "Will you consider medication to help manage your discomfort during the birth?" he asks.

"Yes. Yes, I will," I answer, using his same dry tone that belies the fact that I have never been more passionately sincere in my whole life.

"Good for you," mumbles my nurse under her breath. "Ain't no *reason* for you to be in all that pain." When I hear this I realize that I have just made her shift a whole lot easier.

Every couple of hours a different doctor who is also a professor herds a gaggle of medical students through my room and introduces me as a case study. He summarizes the results of all of my prenatal visits and lists the medications that I have taken in a terse and disjointed monotone, making the whole thing sound like an e. e. cummings poem that didn't quite make his editor's cut. Then he asks his cohort, "So what do we surmise about the fetus in this situation?" and the group responds with the same dumb silence one would get upon querying a flock of sheep.

Finally my nurse breaks in, saying, "Well, look at her. That baby ain't premature, and it ain't underweight neither." As she shakes her head in disgust, I see one of the students standing in the back yawn massively while looking right at me, and without even bothering to hide it.

I suddenly become incensed and it probably shows up on the electrocardiogram to my left. Instantly thrown back fifteen years, I am again a college girl who desperately wants to go to medical school but knows from the start that she doesn't have the money and has no way to get it. I came from a line of women who could catch and pluck an owl, boil it up for the kids and crack the marrow out for the baby, then drink the boil water because that was all that was left. I was the girl who could pull leeches off of herself and wasn't afraid of spiders, snakes, dirt, or the dark. I am suddenly again the girl who, having secured a scholarship that also paid for books, immediately went to the bookstore and bought all the medical texts in addition to the course books that she actually needed.

There these medical students are, on the other side of a heavy iron door that has been locked against me, and instead of glorying within the inner sanctum, they seem to be throwing it all away. I then proceed to wonder indignantly why these little bastards think they are even fit to measure my cervix. My rage awakens a bit of the old me, and in my head I edit the version of these events that I will relate to Bill, and here insert myself yelling: "Write it down, motherfuckers; I'm going to be on the *test*!"

The professor interrupts my internal tirade by announcing, "She has a severe risk for postpartum psychosis and will be observed accordingly," and just like that, he has given voice to something we all suspected but that love and hope had both conspired to keep silent. I perk up, damned interested in what he might say next. Upon receipt of this novel tidbit of information, the students visually scrutinize me anew and appear so nonplussed with my sane presentation that I consider feigning a hallucination in order to validate the professor's point.

While looking around the room in consternation I lock eyes with Clint, who is sitting meekly in his corner chair with his legs crossed. Using the telepathic bond peculiar to married couples, we communicate our mutual recognition of the moment's absurdity and I burst out laughing for the first time in weeks. It then occurs to me that I am feeling the best that I have in months, now that I am perched securely in my wiry little nest of beeping machines.

Immune to both mirth and sorrow, the doctor consults his watch and walks out, with the students trailing behind like the world's lamest paparazzi following the world's most uninteresting celebrity. My anger relents when I suppose that they also have a long night ahead of them. My cooler head then leads me to consider the possibility that dreaming of being a doctor and the reality of navigating medical school are maybe not the same thing, and I also admit that my own demeanor during the past several months hasn't put me in a good position to condemn flat affect when displayed by others.

A surgical nurse enters with what looks like a rolled-up beach towel and proceeds to unroll it across the length of two stainless-steel trays. As he does this, I see that the inside of the sterile cloth is lined with dozens of scalpels, scissors, and various small glinting bladed objects. The assistant leaves and then returns with another towel, identical to the first, and repeats his actions onto two additional trays.

"Oh boy," I observe. "That's a lot of knives."

The nurse looks at me and continues with his task, explaining, "Yeah, this doctor likes to have a second set ready, in case something drops." I am not as comforted as I should be by this assurance that duplicates will be at the ready once the blades start to fly around, but I keep my misgivings to myself as he walks out.

I am gratefully surprised to see my breast-feeding-neutral doctor walk in and announce that she will be attending the birth; I had been told repeatedly to expect any of the doctors who populated my "caregiving team" and was prepared for a virtual stranger, unable to recall even half of the medical characters who had traipsed across the stage of my life during the past nine months.

"I am glad that it's going to be you," I tell her with the trust and affection of a child.

She looks over my chart. "How are you doing?"

"I'm scared," I say, because it is true. I have always been convinced that I will die during childbirth. This is not only because I could never imagine myself as a mother; it is also fueled by my suspicion that this is how my maternal grandmother died. My mother never says much about her own mother, nor her brothers and sisters, except that the ones who survived childhood numbered more than ten. *Diskutere fortiden gir ingenting* (You can't change the past by talking about it).

The doctor stops and looks at me. "If something happens to you," she assures me, "we can have you prepped and in the operating room in forty-five seconds," and I am momentarily enthralled by the idea that there must be another room around the corner with even more numerous—and far more sophisticated—instruments in it than this one.

She then turns to Clint. "That being said, if something happens to you, like you faint, we will kick you to the side and keep going." Clint's mother was a prominent obstetrician in Philadelphia and complicated births were the dinner-table conversation of his childhood, so there is no danger of him fainting, but he nods his acceptance of the scenario described.

She examines my cervix and concludes, "Everything looks good." She adds, "I'll be back after the epidural unless you need me," and then walks out.

A couple of hours go by, during which the blood pressure cuff squeezes my arm encouragingly every twenty minutes and reminds me in happy beeps that I am doing just fine. Then the contractions get really bad and I begin to groan slightly with each one.

"Lord, you sure don't say much," observes my nurse while changing the IV bag.

Taking this as a compliment, I admit, "Well, it wouldn't help to carry on."

"No, it sure don't," she agrees while opening the line that connects the IV to the veins in my arm.

The contractions get much worse and I begin to plead with Clint, quietly begging him to help me in wild-eyed whispers. He stares at me with the calm, friendly face of a Saint Bernard who has just dug you out of the snow and who assures you that a rescue team will be here any minute, and would you like to suck on some ice chips while you wait?

After what seems like hours, a distinguished-looking doctor walks in, accompanied by some sort of lackey, and introduces himself as the anesthesiologist. "Have you ever been treated with ropivacaine before?" squeaks the sidekick while he examines the lower vertebrae of my back, causing me to wonder if this is really something that he expects the average person to know.

After a pause my nurse answers for me. "Probably. Her chart's two inches thick." I begin to suspect that these smart-aleck answers are her hospital trademark, given the practiced way that everyone ignores her.

"Hell, I might be on it right now," I add gamely, my voice shaky with pain, and while looking in her direction. No matter what you say while in the hospital, doctors won't laugh at your jokes. I suppose the official medical school position teaches that no matter how hilarious your patient thinks her condition is, it is not your role as a doctor to guffaw and up the ante, but it still exhausts one to play to such a sober audience.

Fascinated by the fact that they are actively slipping a needle into my spinal cord, I wish desperately to watch the procedure in the same way I had goggled while the nurse poked my arm full of intravenous ports hours earlier.

After a pause the doctor says, "Well done. You are in the right line of work," to what I guess must be an intern who had performed the insertion.

"Yeah, bravo," I add. My thighs start to tingle and I soon feel comfortably numb from the waist down. The pain isn't gone but something has turned the volume knob way, way down.

Presently my doctor comes back and explains how I can use one of the monitors to figure out when a contraction is imminent and then push accordingly, thus adding my voluntary muscular work to the involuntary component. Under her supervision, I do this. For about three hours.

"Okay, new approach," she says brightly. "Did you grow up somewhere where it snowed?"

"Yes," my husband answers for me, "she did."

"Okay, you know how when a car gets stuck in a ditch, you have to rock it out—rock and then push—to get it moving?" she asks.

"In Minnesota, that's just how we park," I reply while panting, and the smile that she gives me is like a hundred-dollar bill that I can stuff into the pocket of my heart.

"Okay, well, that's what we're gonna do, we're going to rock three times, and then you are gonna push," she says, and we try it for a while.

"C'mon, baby, you have a beautiful head, but we wanna see your face," clucks an older nurse as she pats my knee. I synchronize with the arc on the monitor and I push hard, and after I do so, I see the doctor's demeanor change.

She remains calm but stiffens perceptibly and says to the nurse assisting her, "The cord is around his neck. We will do a vacuum extraction." Three people ready a tray of instruments somewhere near my feet, their actions fluid and rapid. The doctor looks into my eyes, intense and serious, and says to me, "This is going to hurt," and I nod my acknowledgment. I briefly notice that she is not wearing earrings, and that I am not wearing earrings, and then everything goes white.

The doctor has attached the suction cup of a ventouse instrument to my son's head, leaned forward, stabilized her weight, and then used all of her might to rip the two of us apart from each other. I hear my own voice shrieking out its bewilderment at finding so many imperfections within a world of limitless potential. When my vision clears, I realize that what I have actually heard is the long-known and already-recognized cry of my new baby.

Now my son and I are side by side, and one team of people is holding and helping him and another team of people is holding and helping me, and we are all covered in my blood, and both of us are just fine. I need do nothing but lie luxuriously and passively marvel at my baby next to me, as it seems as if every single worker in the hospital is busily employed in swabbing the two of us, cleaning us, and checking every single part of both of us again and again. Every detail is being written down and recorded on multiple charts and readouts because we all agree that this data is far too precious to ever be lost or forgotten.

Once my team has stopped my bleeding, they massage a bucketful of now-useless placental chum out of my abdomen while the other team brings my washed and wrapped baby to me for a kiss. "*You* just had a completely healthy nine-pound baby," says a young nurse with a smile.

I smile back at her. "I must be stronger than I look."

"All women are," my doctor adds while scrutinizing the womaniest part of me, improvising a pattern upon which to seam the torn pieces and hem the jagged edges.

Clint is standing next to me, and it is finally his turn to hold and kiss the baby. I look over at my son and I see just enough of my own face in his that I know exactly what he is thinking. He is glad to be born so that he can finally get things started. After Clint puts him back in my arms, he falls asleep and I spend the first of the many, many hours that I will pass during the next months fascinatedly staring at his beautiful face. He sleeps contentedly while my doctor sews, and keeps sewing, and more than ninety minutes goes by. Finally, they pack me with gauze and prepare to leave me with my baby and his father, but not before the blood pressure cuff gives me a hug goodbye, beeps its congratulations, and silently promises to check on me later. The lights are dimmed, and the three of us lie side by side and sleep for hours.

The next days are like a long, happy dream in which I don't have to do anything but lie in bed and periodically testify that I am not psychotic. For reasons known only to the medical establishment, it is crucial under these conditions to establish the patient's sentience with respect to both the day of the week and the identity of our supreme elected official once every six hours. I make a point to proclaim, "Happy Tuesday! Ain't it a grand day for Bush to be in the White House?" toward anyone who walks by wearing a white coat.

On the second day of my stay, the doctor who delivered my son examines my stitches and pronounces my healing to be coming along nicely. After they repack me with gauze and prop me up in the bed, I recommence greedily sucking on my strawberry malt until a bit goes down the wrong way. When I cough forcefully, something gelatinous detaches from inside me and tumbles out, and a bloody stain the size of a dinner plate slowly soaks in between my legs for all to see.

"I don't mean to be a bother," I remark, "but am I supposed to be bleeding this much?"

"There's not a pound of fat on you," answers my doctor. "All that weight was fluid and tissue that you no longer need. It's going to take a while for it all to come out."

As the nurses help me change my bedclothes all over again my doctor adds, "Don't worry. We're all watching you," and after she walks out, I resist hard my temptation to believe that my grandmother might be speaking to me through her.

And so I lie in bed and feel what I don't need come out of me. A steady ooze of bloody, amorphous clots slides out of me for days and with it flows all the guilt and regret and fear that I have carried, and while I sleep, people stronger than I am silently take it all away and dispose of it properly. When I wake, I hold my baby and I think about how he is my second opal that I can forever draw a circle around and point to as being mine.

As we remain in the hospital for another week, the rainy April weather gives way to a dazzling bloom of May sunshine, and the new pattern of our lives starts to emerge. When Clint holds our son I edit a manuscript, or remotely log on to the mass spectrometer, or reject somebody's paper, or sketch out a graph, and we develop a routine that will carry us through the next several years. We pass the baby back and forth, smiling our love to each other during each handoff, and practice doing three things at once. Bill surprises us by visiting the hospital and hugs me for the first and only time in eleven years, and I am amazed to see how easily and willingly he settles into the role of beloved uncle.

All the extra tests performed during our extended hospital stay verify that my difficult pregnancy has ended in a normal, healthy birth. While lying awake during my final night in the hospital, I realize, as I often do, that a problem has eluded me not because it is unsolvable, but because its solution is necessarily unconventional. I decide that I will not be this child's mother. Instead, I will be his father. It is something I know how to do and something that will come naturally to me. I won't think about how weird my thinking is; I will just love him and he will love me and it will just work.

Perhaps this has been a million-plus-year-old experiment that even I couldn't screw up. Perhaps this beautiful little baby at whom I stare anchors me to yet another thing that is greater than myself. Perhaps it will be one of the great privileges of my life to watch him grow and give him what he needs, and let him take my love for granted. Perhaps I can do this. I have help, I have enough money, I have love, I have work, and I have medicine if I need it. Maybe they that sow in tears actually shall reap in joy. Perhaps I can do this too.

9

EVERY LIVING CELL IS essentially just a tiny bag of water. Viewed from this perspective, life (the verb) is little more than the construction and reconstruction of trillions of bags of water. One thing that makes this difficult is that there is not enough water. There will never be enough water for all the cells that could grow. Every living being on the Earth's surface has been conscripted into a never-ending war over a total amount of water that equals less than one-thousandth of one percent of the planet's total.

Trees are at the worst disadvantage because they cannot roam the landscape in search of the water that they need—and because they are large, they need a great deal more than the animals that can move. If you drive across the United States from Miami to Los Angeles on Interstate 10, passing through Louisiana and Texas and Arizona, it might take you three long days, but it will surely teach you the most important fact in all of plant biology: the amount of green that you find at a given location is in direct proportion to the amount of annual rainfall at the site.

If we think of all the water on Earth as an Olympic-sized swimming pool, the amount that's available to plants within the soil would fill less than one soda bottle. Trees require so much water—more than a gallon is needed to build a handful of leaves—that it is tempting to envision the roots as actively sucking the soil. But the reality is quite different: the roots of a tree are absolutely passive. Water flows passively into the roots during the day and passively out of them at night, faithful as the tides of the ocean drawn by the moon. Root tissue functions like a sponge: when placed dry upon spilled milk, it will automatically expand to draw the fluid in. If we then move this full sponge onto dry cement, we will soon see the fluid drawn back out, making a wet spot on the sidewalk. Digging down into any soil, we find it wetter as we travel toward the bedrock.

A mature tree gets most of its water through its taproot, which is the root that extends straight down. Tree roots located near the surface grow laterally to form a netted support structure that prevents the tree from falling over. These shallow roots also leak moisture into the dry soil, especially when the sun is down and the tree's leaves are not actively sweating. Mature maple trees passively redistribute water taken from depth up and out of their shallow roots all night long. The small plants living near these big trees have been shown to rely upon this recycled water for more than half of their needs.

A sapling's life is extremely difficult: 95 percent of the trees that make it to their first birthday will not make it to their second. The average tree seed does not travel far; most maple seedlings take root less than ten feet away from the trunk that supports the very branches from which the seed fell. Thus maple saplings must struggle for light while still in the shade of an adult maple tree that has been successfully capturing and using all of the nutrients in the area for years.

There is, however, one reliable act of parental generosity between the maple and its offspring. Each night beneath the ground, the most precious resource of all—water—moves up from the strong and out toward the weak, such that the sapling might live to fight another day. This water is not everything the sapling requires, but it must help a little, and the sapling needs all the help it can get if one hundred years from now there is still to be a maple tree defending this same plot of land. No parent can make life perfect for its offspring, but we are all moved to provide for them as best we can.

10

DURING THE LAST TEN YEARS we have learned that a tree actually remembers its childhood. Scientists in Norway have been collecting seeds borne by spruce tree "siblings" (that is, half-clones) growing in both cold and warm climates; they have germinated thousands of these seeds under identical conditions and have grown the survivors to maturity within a single forest.

Every spruce does the same thing each autumn: they perform a "bud set," where they stop growing in anticipation of the first frost. The Norwegian scientists have observed that among hundreds of genetically identical trees, grown from seedling to adult side by side in the forest, the trees that had been embryos under a cold climate invariably set their buds two to three weeks earlier than do their counterparts, anticipating a longer, colder winter. All of the trees in their study were identically adapted, but the early bud-setters remembered their cold seedhoods, even though they were consistently ill served by this nostalgia.

We don't know exactly how this memory works. We think it is the sum total of several complex biochemical reactions and interactions. Researchers also don't know exactly how the human memory works. They think it is the sum total of several complex biochemical reactions and interactions.

The year that our son started school, we went to live in Norway for a year. I was a Fulbright Scholar and joined a group trying to figure out what tree memory means for the spruce of today, which experience childhood under one climate, only to be thrust into an adulthood governed by a different climate. Establishing the accuracy of human memory, even within one's own mind, is a difficult scientific proposition. It's much harder to measure memory in an organism with a life span that's more than twice your own.

For our experiments, we exploit the most fundamental difference between plants and animals—namely, that most plant tissues are redundant and flexible: a root can become a stem if need be, and vice versa. The fragmentation of a single embryo can lead to several copies of that plant, each with an identical blueprint of genes. New propagation techniques allow us to answer questions like "Does a tree remember extreme malnutrition experienced during childhood?" by starving one seedling for years while lavishing nutrients upon its identical twin. Such experiments are the only way to find definitive answers; they are deeply repugnant and obviously unethical with human subjects. Plants, in contrast, are fair game.

To start these experiments, I count out a hundred spruce seeds—each one smaller than a sesame seed—and soak them in sterile water for several hours. I sit down and adjust my stool in front of a wall that blows sterile air at me, a gentle mechanical wind. I lose myself for a moment in sentimentality, remembering the young girl that I was twenty years ago, who sat in front of a similar sterile hood within a hospital and searched for her future via painful trial and error. "Everything in front of me is clean, and everything behind me is contaminated," I chant to myself. I rotely line up my tools, not placing anything between them and the wall.

The seeds that I am using were collected by Scandinavian foresters nearly a generation ago, from a conspicuously average tree for which I have pages of description written in Norwegian, in the forced penmanship of 1950. I picture dour blond men in muck boots and wonder if they would be proud of me. I decide that they wouldn't as I mark my reflection within a window of the darkened room: greasy hair, tightly pulled back, and stubborn acne that comes and goes.

I light the Bunsen burner close on my right and set the flame to exactly one inch. It flickers in the airstream and helps to sterilize the air. I drop my right elbow and put an alcohol swab on my left side, instinctively keeping both away from open fire. Using my left hand, I fish out one seed with tweezers and position it. I look through the microscope and turn it flat, remorseful that my hands aren't steadier and swearing off coffee for the third time that day. With my right hand, I make a broad and shallow cut with the scalpel, attempting to peel back the seed coat and expose the embryo.

I press the scalpel to flex the coat, and slide one prong of the tweezers under the embryo. I move the embryo forward, too small to see, and touch the tweezers into a petri dish full of gelatin medium that I spent yesterday cooking and pouring. I close the lid and tape it shut with purple tape—the color that means Tuesday. On the lid of the dish, I circle the area where I dropped the embryo in order to narrow the field that we will search for growth or infection. Under the circle I write a long code in black pen that designates the year, the media batch, the parent tree, and the seed lot. I don't write my initials because we all learned each other's handwriting long ago, just as I can recognize the handwriting of each dead Norwegian forester whom I have never met. My lab mates rib me by not crossing the sevens within codes that they know I will see, poking fun at my American-ness. I check the code that I have written twice for accuracy, whispering it out loud each time. The entire process takes me between two and three minutes. I repeat it exactly one hundred times.

Of the many million seeds dropped on every acre of the Earth's surface each year, less than 5 percent will begin to grow. Of those, only 5 percent will survive to their first birthday. Given these realities, the first and foremost experiment in each tree research study—growing a sapling—is actually an ill-omened fight with near-certain failure. Thus the initial planting of seedlings at the start of a forestry study represents a weary victory won by a stoic researcher with a strong sense of fatalism.

This unique intellectual agony shapes the character of the tree experimentalist and selects for those with a religious devotion to science, patient with overtones of masochism. They neither seek nor win the adoration and glory claimed by nuclear physicists who observe new particles and bluster about the speed of light. I am learning their mind-set as I am learning the substages of embryonic development, and both appeal to me. We plant tiny trees during the night so that they may be baptized with morning dew, and sustain our faith that their measurement will yield knowledge to our scientific heirs, some two hundred years from now.

I gather up the petri dishes and carry them through the basement to the walk-in incubator, where I will leave them in the dark at a temperature of exactly twenty-five degrees Celsius. The incubator is like a humid mausoleum, and I wonder if the faintly moldy smell is real or just my own paranoia. Each embryo rests on a bed of gelatin extracted from thousands of other seeds. This medium will fool my embryos into developing wildly, unrestrained by the seed coat that I removed.

In twenty days I hope to find them splayed out indecently, many times larger than they would naturally be—that is, if a fungal contaminant hasn't gotten to the nutrients first. At that time I will select the healthy embryos and rip them apart slowly, transferring the pieces onto a gelatin made from ridiculous amounts of fertilizer and growth hormones. If I am careful and lucky, I can tear a single embryo into twelve pieces under the microscope. Today I cull the intact embryos from two weeks ago, dismember exactly fifty, and then leave them bleeding cytoplasm in the hope that they'll recover and elongate into something that is green on one end and rootlike on the other. My embryo pieces will spend a month under artificial sunlight, forced to photosynthesize and trying to outrun that damned fungus.

Like Julia Child drawing a finished soufflé out of the same oven into which she inserts an uncooked one, I select a hundred healthy differentiated embryos from the light chamber, swapping for the ones I just dissected. I tuck each of these tiny plantlets into the potting cups I've fashioned out of egg cartons, using one Popsicle stick to make a hole in the soil and another to tuck the seedling under. Occasionally during planting, I notice something odd in one of the samples—some goofy green whorl—and I allow myself ten minutes to stare at it and soak up the pleasure of an unusual moment within this day, week, month of monotony.

I should write down that this one is different, but I don't. I used to note any oddities religiously, but I do it less and less as the years go by. It feels too much like a confidence that I haven't been given permission to share. The first green tissues of a radish seedling are two perfectly heart-shaped, symmetric leaves. In twenty years of growing hundreds of these plants, I have seen exactly two deviants, each with a perfect third leaf—a baffling green triad where there should be only a pair. I think of those two plants often, and they even enter my dreams occasionally, causing me to wonder why I was meant to see them. Being paid to wonder seems like a heavy responsibility at times.

At the end of my day I have arranged exactly one hundred tiny trees into a grid. I take photos, guiltily indulging in forty-five minutes of insipid pop radio (music causes labeling mistakes). The finished seedlings resemble a company of green toy soldiers, and I imagine them as fresh seventeen-year-old World War One recruits, eager to be shipped out with no real idea of what they're getting into. We'll move them into the greenhouse, where they will live in relative bliss for three years and be conscientiously repotted every time their world needs to be a bit bigger.

A collection of the survivors will ultimately be planted in a forest and begin experimental treatment. All our special attention renders it probable that one out of every one thousand embryos that we process will give rise to an adult tree, increasing the odds of success many orders of magnitude over the natural world. In thirty years, perhaps one of the plants before me will bear seed and help give answers to the questions that we ask today. That is, if the university doesn't cut down our forest in order to build a dorm, a day care center, or a fast-food courtyard.

At eleven-thirty in the evening I call Bill and he picks up after two rings. "All quiet on the western front," I tell him, and he understands. It is morning where he is, and I have just woken him up.

"Okay, I'll be there in a bit." Then he asks me, "Did you soak the Popsicle sticks?"

"What?" I ask, pretending not to understand.

"Did you soak the *fucking Popsicle sticks* in bleach this time?"

"Yes," I lie, and he snorts, unconvinced.

"Yes," I insist, "I soaked them; I soaked the embryos and I drank a glass of it before I started."

He continues, "Because a year from now, when we're up to our eyeballs in contamination, this whole thing will somehow lose its poetry."

"Well, hopefully it won't take that long," I retort, "because we're out of *fucking bleach,*" and we both laugh.

* * *

We laughed because it was a joke: Bill was not actually on his way to meet me that night because he was on the other side of the world.

During the years after my son was born, being a scientist became easier, although I am still not quite sure why. It surprised me, because even though I hadn't changed the way I designed my experiments or talked about my ideas, the establishment changed the way that it thought about me. I won contracts, not only from the NSF but from the Department of Energy and the National Institutes of Health. Private donors such as the Mellon Foundation and the Seaver Foundation found me worthy of support. This additional funding didn't make the lab rich, but for the first time we could build new instruments, replace broken parts, and sleep in decent hotels while we traveled; best of all, I could chart out Bill's salary for a year at a time instead of from month to month.

Once I wasn't stressed to distraction about our survival, my patience returned and I rediscovered my love of teaching. The combination of freedom and love is a potent one, and it made me more productive than ever. I summed up my ideas about plant development within longer works, formatted as whole chapters, that allowed for the necessary detail. I started to win awards for these ideas once they were fully expressed: first the Young Scientist Award of the Geological Society of America and then the Macelwane Medal of the American Geophysical Union, which made my tenure decision a no-brainer in 2006. Encouraged, I started to take even bigger risks: I applied to do the spruce experiments in Norway—I wanted to learn to plant tree seedlings. I wanted to know what tree memory was all about.

While I lived in Norway, Bill stayed back at home, running the lab. Clint's easy charm and rare mathematical gift have brought him several standing job offers over the years; he accepted one of them and we moved near Oslo together, where we enrolled our son in a Norwegian kindergarten.

I have always felt at home in the glittering fjordlands of eastern Norway. There, no one ever perceives me as cold or standoffish; I can just be who I am. I love to speak Norwegian, which is a terse language in which every word counts double and the whole meaning can turn on the lilt of just one vowel. I love the dark, snowy nights of winter and the endless pastel days of summer. I love to walk through spruce needles and pick berries and eat fish and potatoes seven days a week.

During that year, I loved everything about living in Norway, except for how much I missed Bill. But deep down we both knew that the separation was good for us: we were getting older, and I was raising a family. Convention and circumstances dictated that we should act more like coworkers and less like twelve-year-old fraternal twins.

* * *

Halfway through my year of living in Norway, I sent Bill a text: “I am thinking of you.”

As soon as I sent it, it appeared in my outbox as the last in a long chain of identical unanswered texts that I had been sending daily for three weeks, interspersed with choruses of “I hope you are okay.”

I hadn’t heard from Bill in more than a month. I knew he wasn’t lost, although I felt as if I was. Four weeks earlier I had woken up to the following e-mail from him: “Hey I just got word that my dad died today. Guess I’m going to California. I’ll shut down the mass spectrometer before I go.” I immediately began to text the above mantras, embellished and frequently at first, settling into once daily over time. I never heard anything back.

For weeks after his father’s death, Bill’s e-mails had stopped and my string of unanswered texts left a void in my life. I worked my usual hours but often caught myself staring unproductively at the wall, questioning for the first time why I was doing this science, and finally realizing that it was pointless to do it alone.

Though I heard nothing from him, I knew Bill well enough to know exactly what he was doing. He was working hard each night from 7:00 p.m. to 7:00 a.m., seeing and talking to no one. This was his usual pattern when he was "in a funk," usually post-migraine, and everyone in the lab knew to leave him alone until it passed.

This funk had dragged on, however, and I couldn't help but imagine what his week of mourning in California had been like. How the coming of the dusk dissolves the laces of the splint that holds you together during the day, and the desperate sadness that follows can be anesthetized only with sleep. The heaviness of opening your eyes the next morning when you realize that you've begun another day of grief, so pervasive that it removes even the taste from the food that you eat. I knew that when someone you love has died, you feel that you have also. And I knew that there was nothing that I, or anybody else, could do to fix it.

I kept texting daily and never received a reply. Finally I sent Bill an e-mail: "Hey, let's you and me go into the field. Ireland. You always love Ireland. I bought you a ticket, it's attached as a PDF. Your dad was a good guy. He was good to your mother, and he was faithful to her. He loved you kids, and he was home with you every night. He didn't drink, and he didn't hit people. That's what he gave you. That's what you got, and it's a lot. That's what we got. It's more than what some people get, and maybe it's more than what most people get. And as of now it has to be enough. You land before I do, but the rental car is in my name, so wait for me."

There was so much that I wanted to add, but I didn't. I wanted to venture that Bill had been his dad's baby and his favorite, a final son who came to him late in his life, bringing with him one last precious chance to enjoy childhood by proxy. I wanted to tell Bill that he embodied the happy ending of his father's life, that he was the comforting unspoken punch line to the dark genocide jokes that he told, and that his very flesh constituted a triumph over injustice and murder. I wanted to tell Bill that he was his father's heart and prize, a strong, sinewy boy whom the world couldn't maim, smart and lithe even underground. I wanted to reassure him that, like his dad, he would survive, but I didn't know how. I wrote what I wrote, pressed "send," and packed my gear.

I flew to Ireland and walked off the plane into Shannon Airport to find Bill standing beside three huge duffel bags that had been stuffed full of tools and then strapped with duct tape. "Good God, have you run away from home?" I asked him, smiling. "What exactly did you think we were going to sample on this trip? The ocean floor?"

"I didn't know what the hell to think," Bill answered. "Your e-mail didn't say a damn thing. I couldn't take chances, seeing as this is a third-world country. So I brought everything." Bill's manner was somewhat subdued and he looked tired but otherwise fine. *He will survive this,* I thought; *we both will.*

I did have a plan, but only sort of. First, we went to the airport shop and I bought two packages of every different kind of candy that they had for sale. "Provisions," I explained. At the rental car desk, the man behind the counter asked us if we were married. "Maybe," I hedged. "Will it affect our rate?" He explained that the additional driver fee is waived when it is applied to a spouse. "Well, then, yes, I seem to remember that we are married. Isn't that right, dear?" I prompted Bill with my elbow. I saw Bill's face blanch and he looked to be stifling the urge to vomit. I smiled with satisfaction.

The clerk asked us if we had our own car insurance. "Yes," I answered. He then asked if I wanted to supplement it with additional insurance, and I automatically answered, "Yes." He asked if we wanted to fully cover both the car and the—and I cut him off with another "Yes."

The clerk looked at me quizzically. "It's bleedin' expensive, you know," he told us, possibly confused by the fact that only moments ago I'd been willing to pretend holy matrimony in order to save five bucks a day.

"It's not as expensive as some other things," I answered mysteriously as I signed and initialed the stack of papers.

Finally, the clerk described the car and pointed us on our way. "Right, then. The petrol is prepaid; the cleaning is prepaid; the vehicle is insured; the drivers are both insured; any damage to any other vehicle is insured; if anything happens—"

"We walk away." I finished his sentence for him. "We just walk away."

"Yes," affirmed the clerk, but he looked troubled as he handed me the keys.

"'Bleedin' expensive, you know,'" Bill mimicked as we walked out, looking for the car. "Why is everything so bloody-bleeding over here?"

I launched into a discourse on the gradual contraction of medieval oaths invoking the Virgin Mother Mary's menstrual blood and the seepage of Christ's wounds, astounded to find any use for my college study of medieval literature. I was driving, and eventually we settled into a comfortable silence, watching a foreign world go by from the wrong side of the road. We'd been to Ireland many times before: the massive layered cliffs of coal along its western coast are a wonderful place to teach students how to identify and map fossil-bearing rocks. On this trip, I was doing the driving—for a change, I was going to be the strong one, taking care of everything.

"Let's go through Limerick and not around it, what do you think?" I asked Bill. He shrugged to signify that he didn't care. I took the roundabout off of the N18 and eased onto Ennis Road, then headed south toward the Shannon Bridge.

"Ughhhuh!" Bill suddenly made a retching noise and spit a huge wad of tar out the window and into the River Shannon. "Whatever that was, it just threw up in my mouth." He pointed to a package of black candies that had turned out to be pungent gelatinous licorice rolled in salt and not sugar. "Zounds!" he added, referencing our recent conversation.

"It's an acquired taste," I remarked, giggling at his discomfort. Bill didn't laugh, but his eyes lightened and I thought I saw his misery leave him for a moment. "Do you want me to throw the rest of it at this policeman, or bobby, or whatever he is?" I offered, rolling down my window.

"Naw." Bill slumped in his seat. "I'll probably eat the rest of it later." We turned north on O'Connell Avenue and headed for the Milk Market district. "What are we doing here?" asked Bill, rather philosophically, I thought.

"We're looking for leprechauns," I answered pensively. "Keep your damn eyes open." I was getting lost, confused by streets with names like "Sráid Eibhlín" and "Seansráid and Chláir," but I didn't care. I wasn't trying to find anything; I was waiting for something to happen.

The roads narrowed and I drove on, taking whichever claustrophobic turn looked to lead down the most obscure alleyway. I turned my head toward Bill and was about to wonder aloud what an "Arms" was anyway, having passed by the Johnsgate Arms, the Palmerstown Arms, and several others, when I heard *"Bam!"* and felt a bone-rattling crack go through the car.

I slammed on the brakes, wondering why anyone in this tame neighborhood would try to bash our car window in with a baseball bat. My hands were still shaking on the wheel when I looked to the right and saw only the silhouette of Bill's head, lit from behind with a glowing halo made by the spider-web pattern in his passenger's-side window. We scrambled numbly out of the car on my side, and Bill lumbered around to the other side of the car in order to see what had happened. I sat down on the curb and tried to still my nerves.

"God, these car accidents are a lot less fun than they used to be," I told Bill, and he agreed.

Unable to accurately judge the position of the vehicle from the right side of the road, I had been driving too close to the curbside. This had gotten worse until I passed near enough to a streetlight such that it snapped the passenger's-side mirror clean off and smashed it into Bill's window.

"Well, you've made a holy show of yourselves," remarked a man in an apron, coming out of a nearby pub with a few other people who had heard the glass shatter. He whistled at the car. "Now, that will be dear to put right."

Bill saw an opportunity for diplomacy in action. "We're Americans," he clarified. "Our plan is to just walk away."

"And what in the heavens would you be after in County Clare?" asked a rather short, jovial, and particularly Irish-looking bystander.

Bill looked the guy up and down and answered, "I think we were looking for you." Bill then turned, picked up the broken mirror, and unceremoniously tossed it into the trunk of our car. He rummaged through one of the duffel bags and pulled out a big roll of clear strapping tape.

An elderly gentleman from the pub addressed Bill. "If it'd had five degree warmer, down with the windows and off would have been your head!" He laughed at his observation, as did his companions. "She'd be trying to kill you, by the look of it . . . ," he added, shaking his head and motioning toward me.

"I know," agreed Bill, "and the sad thing is, we only just got married this morning."

Shaken and embarrassed, I declined the spectators' cheerful urging that we should come in and have a pint or two. Bill set about to secure the broken window, carefully layering tape all over the outside and then layering it all over the inside as well, and I helped him by unrolling the lengths that he specified. Gradually I began to feel normal again, and it seemed that Bill had found yet more of his old self. James and John in a boat with their father, I thought, mending the nets and waiting to be called. We would make everything hold together, even if it was never to be the same.

"You want to drive?" I asked Bill sheepishly as we put on the final touches.

"Naw," he said, "you're doing an excellent job." He slid into the car deftly, stabilizing our jerry-rigged window with one hand. "But let's get the hell out of the city," he suggested. "I need some green."

Traveling southwest on the N21, we lived anew our very first impression of Ireland from five years earlier: the greenest place in the world. Ireland is so saturated with green that it is the things that are not green that catch one's eye. The roads, walls, shorelines, and even sheep seem to have been placed as contrast, strategically positioned to organize a vast expanse of green into its billion distinct subshades of light green, dark green, yellow-green, green-yellow, blue-green, gray-green, and green-green. In Ireland, you can bask in the fact that you are benevolently outnumbered by these first and better life forms. Standing within a peat bog in Dingle, you can't help wondering what Ireland was like before you and the other primates scrambled up upon its shores. When viewed from space, did it glow like a furry emerald within a sea of blue, the terrestrial equivalent of a massive marine plankton bloom?

We arrived at the Phoenix—a B-and-B/organic farm where we usually camped—and its proprietors, Lorna and Billy, greeted us warmly, as usual. When they shook their heads over "langers acting the maggot" in Limerick, we weren't sure that they didn't mean us.

"Will you take any tea?" Lorna asked us. "I know youse hate to get hiking until it's fairly good and pouring out." We sat down, drank a pot of tea, slathered a loaf of soda bread with butter and currant jam, and ate it. After that, we sat and looked out the window, waiting for a productive restlessness to set in.

"Well," said Bill at length, "my boots aren't wet."

"That's fixable," I said, taking his cue. We geared up to go out for the day. By that time, every field trip began with our now-unspoken habit of driving to high ground and then parking and hiking up to the highest point we can find. Once there, we stand and look as far as the eye can see and wait for an idea to come to us. Because all the best-laid plans in the world can be rewritten into something better from the right perch, we've stopped making detailed plans in advance, trusting instead that only from the top can we really see the way.

Bill stared at the horizon, but not in the serene and contented way that he usually did when freed within wide-open spaces. Instead, he wore a heaviness, worn out from having carried his grief halfway around the world. We stood side by side and looked out.

At length I spoke. "It's hard to believe your dad is gone," I observed simply; this had been my first reaction to his loss.

"Yeah, I know," he agreed. "It was a surprise," he admitted. "I mean, who the hell would have expected a ninety-seven-year-old man to just up and die?" Bill's dad had indeed been only three years from his one-hundredth birthday when he shocked everyone by waking up dead one morning.

"We never suspected it, but it turns out that he was just old as shit," he added. I remarked that after a man reaches ninety-five and hasn't died yet, the people around him become lulled into believing that he never will. Right up to the end, Bill's dad had continued to work in his home studio, doggedly editing the massive pile of footage that represented his sixty-year career as a filmmaker.

"Yeah, but was it a stroke, or a heart attack, or what?" I prodded gently.

"Who knows? Who cares?" Bill answered listlessly. "They don't do autopsies on ninety-seven-year-old bodies."

"I just picture him barging into Heaven," I offered, "pushing right past the place where you get all the answers to the big questions, learn why there is so much suffering in the world and why we are here and all that, he just beelines to some corner, unrolls a length of rusty chicken wire, and stakes it down with old coat hangers so that he can start planting tomatoes."

"Oh, I'm not worried about him," returned Bill. "He's gone. It's not any more complicated than that. Honestly, if I admit it, it's me that I feel bad for." He walked away from me and looked out toward the south. "There's nothing like having a parent die to make you realize how alone you are in the world," he added.

I kneeled. A few meters away, Bill's back was bent but his body was standing. There were so many things that I wanted to say. I wanted to tell Bill that he wasn't alone and that he never would be. I wanted to make him know that he had friends in this world tied to him by something stronger than blood, ties that could never fade or dissolve. That he would never be hungry or cold or motherless while I still drew breath. That he didn't need two hands, or a street address, or clean lungs, or social grace, or a happy disposition to be precious and irreplaceable. That no matter what our future held, my first task would always be to kick a hole in the world and make a space for him where he could safely be his eccentric self.

Most of all, I wanted to wrench Death off and send it back where it came from; it had reaped enough hurt from him for now and would have to be satisfied with an IOU for the future. The unfortunate fact was that I didn't know how to say any of those things out loud, and so I only rubbed the snot streaming from my nose and thought them to myself.

When I reached down to wipe my hands on the moss, I was surprised by how comforting the soft, spongy turf felt. My knees had sunk into the top layers of the sod and the water that had squeezed out was pooling and soaking me through. I reached back down and ripped up whole handfuls of moss, rubbing it between my hands to "wash them dirty," as we liked to say. I looked at the debris that stuck, and up close I saw what looked like tiny feathers, Kelly green on the topside, lemon-green on the underside, and with streaks of faint red along some of the edges. A pigment for every sunbeam no matter how wan, I thought as I looked up at the clouds.

The rain had amped up a notch, changing from a drizzle into a steady leak out of the sky. As I stood up, I felt the chill travel up my legs and settle into my bones; beneath my woolen long underwear I could feel water running down my legs and soaking my socks from above. I knew that I wouldn't be putting on thoroughly dry clothes again until we had left the country. When you're wet and cold and stumbling in the muck, the plants around you appear smug in their superiority, not only tolerating the miserable weather but thriving on it.

"Yeah, you love this shit," I sneered to a clump of moss in front of me and stomped directly upon a little hillock like a petulant child, frustrated over something unrelated but unable to conceive why. The moss flexed downward unharmed, disappearing underneath a pool of clean, clear water, and then sprang back when I removed my foot, not even preserving the impression of my boot. I sighed. "You win, asshole," I conceded, and felt depressed. I considered it, stomped on it again only to achieve the same effect, and then kicked it, and it did the same thing.

"Riverdance?" Bill had turned around and was watching me with bland interest.

"Do you have any twenty-five-milliliter vials with you?" I asked.

"Only three hundred," he answered. "The mother lode is back in the gray duffel bag."

"You know . . . because these things look just as fat and happy as the stuff at lower ground . . ."

I was talking about the moss, and Bill picked up on it immediately and finished my thought for me: ". . . even though water should be more available down low, near the streambed."

"It's a living ShamWow," I said while working my foot up and down, showing him the way the water pooled when the plants were condensed.

"But does it hold as much water here as it could if it were growing in the lowlands?" Bill asked while staring at the horizon, and we both knew that we had found our question for the day, and possibly for the trip.

Conventional wisdom holds that plants sit on the landscape and wait for water, wait for sun, wait for spring, wait for everything to fall into place before they take their cue to grow. If plants were indeed the passive agents that they were purported to be, water would run right through what was obviously a porous substrate and pool in the lowlands, and we'd have seen conspicuously more green down low. But what if it was the moss itself that was keeping the high ground so mushy, hanging on to water that would otherwise have run down the hill, spreading the wet for its own purposes?

What if this moss had moved into an area, deemed it not wet enough, and proceeded to change this high ground into the soggy mess it preferred, causing what was previously heterogeneous to evolve into a uniformly green expanse? What if the landscape wasn't setting the stage for plants, but the plants were setting their own stage, green begetting green begetting green? What if it couldn't be stomped down, beaten back, or dried up? What if we were slipping, sliding, and stumbling across something stronger and steadier than ourselves?

"Carbon isotopes in leaves should give us water status; we can directly compare the values in the upland with the lowland moss," I said, summarizing my hypothesis, and started digging in my backpack for Atherton et al.'s *Mosses and Liverworts of Britain and Ireland,* an eight-hundred-page behemoth that categorizes and describes the salient characteristics of the approximately eight hundred species of British and Irish bryophytes. I opened it and began to read, hunched over so that my body blocked some of the spitting rain.

The introduction told me that I would need to magnify each leaf, which was about as big as a fingernail clipping, by at least ten and maybe twenty times in order to see the features that identify the different species. "We have magnifying lenses, right?" I asked Bill, and added, "Atherfuck also says that mosses are best identified when wet."

"Well, we should be okay on that one," said Bill as he wrung the water out of his fingerless gloves ("I'm so sick of wasting money on fingered gloves," he had explained while purchasing them the year before at REI).

We settled onto our knees and began to take inventory of the species near us. After two hours, we were pretty sure that we'd found *Brachythecium* thanks to its furry, leggy appearance up close ("Upon 20× magnification the fronds resemble Oscar the Grouch's pubic hair," Bill wrote in our field notes using his careful script). We were only partly convinced of its species (*rutabulum* was the front-runner), and so we settled upon *Brachythecium oscarpubes* for the time being.

Members of the *Sphagnaceae* family were not hard to find, given the rich red pageantry of its incipient leaves, though for the life of us we couldn't place the species. After a long digression as to whether we should include the puffballs of *Polytrichum commune* ("Because they're so pretty," I argued scientifically), we agreed to limit ourselves to *Brachythecium* and *Sphagnum,* considering it likely that we would also find these two genera in the lowlands.

Bill was writing everything down in detail. "How many of each?" he asked, mentally calculating how the contents of each vial would be conflated into three separate analyses on the mass spectrometer for carbon isotope composition. He answered his own question as he quickly recounted the number of vials that we had with us: "I guess no more than a hundred and fifty."

"Let's just sample until it gets dark, and see what we get done," I said while carefully noting our exact location on the topographic map and verifying it against our GPS. We negotiated a labeling code that incorporated the date, site, species, number, and the collector responsible, and then took out our tweezers and got to work. "Everything we've ever done and read tells us that the individual variability is high, so the more of these we can get home, the closer we'll get to measuring the site average," I mused.

"If there is such a thing as a site average isotope value." Bill hit upon the ultimate sticking point of our study.

By the time we'd collected twenty *Sphagnum,* we'd found our groove: first I proposed a tissue for collection, then Bill confirmed that it belonged to a distinct and identifiable individual, then I photographed the plant against a scale card while Bill wrote down anything notable, then I picked and placed the sample in a tube and capped it, and then Bill labeled it and set it down in order. At the end of this process, we retraced each step for accuracy and I reread the code on the label while Bill verified it against its entry within the field notes.

I believed that photographing each individual sample was over the top, but I let Bill have his way and gave thanks to the digital age that was saving us thousands of dollars compared with what we'd spent over the years in developing roll after roll of film cataloging identical-looking leaves.

We crouched in the wet mulch so close together that the tops of our heads touched now and again. "I want you to know that I feel a lot better now," Bill said as he worked. After taking a deep breath he added, "Which is surprising considering the large number of open sores covering my scalp."

We worked until our shadows grew long and the dusk began to fall. We gathered up the vials that we had filled and bundled them into ziplock bags, carefully labeling them in batches. We drove back to the farm and peeled off the outer layers of our soggy clothes, and then we sat by the fire, steaming in our long underwear until late into the night.

We repeated this collecting routine at seven more sites, four located upland and four in the lowlands. When we packed up to leave the country we had more than one thousand hand-labeled vials, each containing a single leaf that had been identified, described, photographed, and cataloged.

"If it's moss youse after we'll get straight to making more and so bring you back," Billy told us as he sent us out at four in the morning with a big hug, and we got into the car and drove off toward our morning flight.

Bill drove and I dozed fitfully against the window, nagged by guilt for failing to entertain him during the long, dark drive. Upon arriving in the car rental parking lot at the airport, we retrieved the broken side-view mirror from the trunk, taped the keys onto it, and heaved the whole thing into the after-hours key return. We got a bus to the terminal, checked our bags, printed our boarding passes, and headed through security.

The moss samples were in our backpacks. We had learned long ago never to check samples unless it was absolutely unavoidable: however remote the possibility that an airline might lose our baggage, it was too much to risk where samples were concerned. When we placed our bags onto the x-ray conveyor belt, the glass vials tinkled. We walked shoeless and docile to the other side of the checkpoint, only to find a security agent waiting for us.

"Now, then, you'll be having a permit for these?" She had opened our backpacks and was handling our samples as if they were wads of garbage that she had pulled out of a trash compactor.

Oh shit. Permits, I thought. We didn't have one, and I wasn't entirely sure that we needed one just to get them to Norway. I should have looked it up before the trip, instead of worrying about Bill. I wracked my brain for a believable lie, or a funny story, or anything that might get her to hand the samples back to us.

Bill was always direct and honest when answering the questions posed by people in uniform, and it never failed to impress me. "No permits needed, because they're not endangered plants. We're scientists; they're just for our collection," he explained calmly.

The agent had opened one of our ziplock bags and was roughly sifting through the vials with her hand. A couple of them bounced out of the bag and fell onto the ground. She pulled a single vial out of the bag, held it up to the light, and shook it; she unscrewed the top of the vial and turned it upside down. It was like watching someone shake a baby. I stretched out my arms, mutely hoping that I could appeal to her in terms of basic female sympathy and she'd pass them back so that I could cradle them, settle them back into place, and shush them to sleep.

"Nope," she chirped. "Biological samples don't leave the country without a permit." She scooped up the entire lot and chucked it into the refuse barrel with one motion. I looked at the pile of discarded items that had been abandoned at the last moment. There were bottles of drinking water and hairspray, Swiss Army knives and matches, an open container of applesauce, and a big pile of tiny glass bottles, each of them covered in meticulous handwriting and containing a speck of precious green. Sixty hours of our lives were also buried in that pile, and possibly also the answer to an important scientific question, I thought. Bill pulled out his camera, leaned into the barrel, took a photo, and walked away.

We trudged over to Bill's gate, walking on the backs of our shoes. In an hour he would be returning to the United States, and my flight wouldn't leave for Norway until later in the day. We sat down to wait, and Bill began sifting through his address book and scribbling down 1-800 numbers. He checked his watch and said, "I get to Newark at nine a.m. East Coast time. I'll call the USDA when I touch down and see what we need to get a permit to bring plants out of Ireland."

I sat stewing in defeat. *The samples we've lost over not having permits,* I thought. *The time we've saved by not applying for them,* I countered. *When will I learn?* I asked myself.

Bill interrupted my internal dialogue by looking at me meaningfully and saying, "It'll never be really lost, you know; we wrote everything down. We'll start over. When you think about it, we got a lot done on this trip." I nodded, and before long his zone was called to board, and for the second time that day something was pulled away from me that I didn't want to let go of.

I watched Bill's plane push back and taxi off, and I thought how the more important something was in my life, the more likely it was to go unsaid. Then I pulled out my vegetation map of southwestern Ireland, lined it up with the topographic map, and systematically planned out where we could find more moss.

* * *

Bill would forever after refer to that trip as "the Wake," whereas I dubbed it "the Honeymoon," and we took to reenacting its climax at least once a year. Whenever we got a new recruit to the lab, his or her first task was to label empty vials, hundreds of them. We'd explain that this was a necessary preparation for a large-scale collection we had scheduled and give directions for a long and complicated alphanumeric code, rich with Greek letters and nonsequential numbers, to be inscribed on each vial in pen, along with the order of production.

After a day of steady labor on the part of the newbie, we'd hold a summit and either Bill or I would play Good Cop and the other would play Bad Cop (we traded off). The meeting would start out with our asking the newbie how he or she had liked the task and whether this sort of work was tolerable. It would then slowly morph into a discussion of the upcoming sample collection and the rationale behind its purpose.

Little by little, Bad Cop would become more and more pessimistic as to whether the proposed collection would test the hypothesis after all. Good Cop would resist this logic at first, urging Bad Cop to consider the fact that the newbie had put so many long hours into the preparation. Even so, Bad Cop just couldn't let go of the nagging realization that this approach wasn't going to yield an answer, and finally Good Cop had no recourse but to agree that starting over was as unavoidable as it was necessary. At this point, Bad Cop soberly gathered up the vials and dumped them en masse into a lab waste receptacle. The Cops exchanged a knowing look, and Bad Cop trudged off without satisfaction, leaving Good Cop to observe the newbie's reaction.

Any sign that the newbie regarded his or her time as of any value whatsoever was a bad omen, and the loss of so many hours' work was a telling trial of this principle. As a corollary, any recognition of futility was perhaps worse. There are two ways to deal with a major setback: one is to pause, take a deep breath, clear your mind and go home, distract yourself for the evening, and come back fresh the next day to start over. The other is to immediately resubmerge, put your head under and dive to the bottom, work an hour longer than you did last night, and stay in the moment of what went wrong. While the first way is a good path toward adequacy, it is the second way that leads to important discoveries.

One year I played Bad Cop but forgot my reading glasses and so returned early to the melee. Our newbie, named Josh, was busily digging his vials out of the refuse bin, separating each one carefully from the used gloves and other trash. I asked him what he was doing and he said, "I just feel bad that I wasted all these vials and stuff. I thought I could unscrew the caps and save them, and they could be extras or something." As he continued with his task, I caught Bill's eye and we smiled at each other, knowing that we'd identified yet another sure winner.

11

LIKE MOST PEOPLE, my son has a particular tree that figures prominently in his childhood. It is a foxtail palm (*Wodyetia bifurcata*) that sways amiably in the wind through the endless months of Hawaiian summer. It stands just a few feet from our back door, and every afternoon, my son spends about thirty minutes hitting it as hard as he can with a baseball bat.

He has done this for years, though not always with a bat. The scarring on the trunk starts down low and then proceeds upward, tracking my son's growth. When he was four, he would use all his tiny strength to bring our sledgehammer against it over and over, while pretending to be Thor. This was followed by a period when an old golf club served, and when the dog quickly learned to avoid the general area. My son's recent baseball obsession has furnished an agreeable cover story: he now bashes the tree exactly one hundred times a day in order to "strengthen his swing." The wood-on-wood quality of this new approach represents an interesting parity from my perspective, and I freely admit that I don't feel inclined to intervene.

He's not hurting the palm tree; if you compare its crown with the next one over, you'll see that they both have a similar amount of healthy green fronds at the top. It also flowers and bears fruit the same way that it always has, just as well as or better than any other palm tree in the neighborhood. My son has never shown the least interest in walloping any other living thing, and as it doesn't seem to be so much about hitting as it does about ritualistic noisemaking, the beating of this living drum has become the rhythm of our lives.

Every day I sit at our kitchen table and write while my son works over the palm tree.

In 2008 we moved to Hawaii, lured not so much by the gorgeous weather and lush vegetation as by a promise (in writing!) of 8.6 months of guaranteed salary per year for Bill "in perpetuity" from the University of Hawaii. That still leaves fourteen weeks of his salary that I have to beg from government contracts each year—but hey, they wouldn't want me to get lazy or anything.

Since moving to Hawaii, I've learned that palm trees are not really trees: they are something different. Inside their trunks you won't find hard wood growing outward, new tissue added ring by ring. Instead you'll find a jumble of spongy tissue, scattered instead of arranged. This lack of conventional structure is what gives the palm its flexibility and makes it supremely adapted to my son's favorite hobby, as well as to the gentle island breezes that periodically coalesce into ruthless hurricanes.

There are thousands of different palm species, and they all belong to the *Arecaceae* family. The *Arecaceae* are important because they were the first plant family to evolve as "monocots" about a hundred million years ago. The first real leaf of a monocot is a single blade, not a double sprout as in the "dicot" plants that came before. My son's palm "tree" is much more closely related to the blades of grass within the lawn beneath it than it is to the monkeypod tree beside it.

The very earliest monocots soon evolved into grasses, and grasslands eventually spread across the vast areas of the Earth where it's just a little too wet to be a desert and still a little too dry to be a forest. With some breeding help from humans, grasses were evolved into grains. And today, just three monocot species—rice, corn, and wheat—provide the ultimate sustenance for seven billion people.

My son is not me: he is something different. He is naturally cheerful and confident, and he has inherited his father's emotional stability, whereas I tend toward nervousness and brooding. He views the world as a racecar and assumes that he should be driving, while I have always focused upon not getting run over. Indeed, he is happy with what he is and does not question it—at least not yet—whereas I will forever be stuck in the in-between.

I am neither short nor tall, and neither pretty nor plain. My hair was never quite blonde; nor was it brunette either, and lately it has become only sort of gray. Even my eyes are neither green nor brown—everything about me is hazel. While I am too impulsive and aggressive to think of myself as a proper woman, I will also never fully shake this dull, false belief that I am something less than a man.

Because we are so different, it took me a long time to figure out what my son had to do with me. I am still learning the answer. I had worked so hard for so many years trying to *make* my life into something that it was a surprise to see all the truly valuable pieces simply fall from the sky undeserved. I used to pray to be made stronger; now I pray to be made grateful.

Every kiss that I give my child heals one that I had ached for but was not given—indeed, it has turned out to be the only thing that ever could. Before my son was born, I anguished over whether I would be able to love him. Now I worry that my love is too vast for him to understand. He needs to know a mother's love, and here I am, impotent to express the fullness of it. I realize now that my son was the end of a waiting that I didn't even know I was doing. That he was both impossible and inevitable. That I have been given one chance to be someone's mother. Yes, I am his mother—I can say that now—for only after I released myself from my own expectations of motherhood did I realize that they were something I could fulfill.

Life is funny that way. While my son was growing inside me, I did the breathing for both of us. Now I go to his little school pageants, sit in the audience, and see only his face, though the stage is filled with children. I breathe in heavily after he sings each stanza, convinced that I can oxygenate his body from a distance, just by the sheer force of my love. He is growing up, and I have to let him go a little more each day. I have learned that raising a child is essentially one long, slow agony of letting go. I am comforted by my suspicion that all my private maternal ecstasy is really nothing more than what every mother feels for her son.

And for a daughter? I'd like to think that it's also felt for her, but this will not be mine to know. Being a daughter was so difficult for both my mother and me; maybe our line needs to skip a generation in order to extinguish the cycle such that it cannot be repeated. So I've set my heart on a granddaughter—as always, my greed for love is unreasonably premature. Based on my projections, there's more than a small chance that I'll die before she's born, particularly if our line continues to skip or bifurcate. And perhaps this is the way that it was meant to be, for me anyway.

Nevertheless, here on this sunny day, I can't resist my temptation to put a message in a bottle: Somebody remember. Somebody someday find my granddaughter and tell her. Tell her about the day that one of her grandmothers sat looking out of her kitchen window with a pen in her hand. Tell her that her grandmother didn't see the dirty dishes or the dust on the windowsill because she was busy deciding. Tell her that in the end, she decided to go ahead and love her granddaughter several decades too early. Tell her about the day that her grandmother sat in a sunbeam and dreamed of her to the soundtrack of a tree being flogged.

12

AS I WALKED INTO THE LABORATORY, the look on Bill's face told me two things: first, that he'd been up all night, and second, that today was going to be a good day.

"Where have you been? It's seven-fucking-thirty." Bill's version of "good morning" has changed very little over the past twenty years. When he was living in his car, it was the suffocating heat of the Atlanta sunrise that drove him into the lab at this early hour. These days if he's here before ten, it's because something happened last night that was too good to leave. And on that particular morning, he had called me.

"Sleep is for the weak!" I barked out. "What's up?"

"It's C-6," he answered. "That little bastard is doing it again."

He led me past the growth experiments, where eighty radish plants had been growing for twenty-one days under precisely controlled levels of light and moisture, within chambers of perfectly still air. One of the biggest ironies of C-6 came from our assumption that we weren't going to see anything interesting. In fact, the experiment was designed to let us measure something we couldn't see.

For any given plant, the part that can be seen is only about half of the whole organism. The roots that live under the soil have nothing in common with the green foliage that extends above the surface; they are as different as your heart is from your lungs and are likewise adapted for two completely different purposes. Aboveground plant tissue works to capture light and gases from the atmosphere, which are converted to sugars within the leaves. Belowground tissue strives to absorb water and the rich nutrients that are dissolved within this water, in order to further build sugars up into proteins. Green stem gracefully morphs into brown root at the soil's surface, and somewhere inside that interface important decisions are made. If both ends of the plant succeed, there is then the question of what to do with that day's winnings. The making of sugars, starches, oils, and proteins is all possible, but which ones of these should be constructed?

Upon gaining new resources, a plant may perform one of four actions: it will either grow, repair, defend, or reproduce itself. It can also delay its choice indefinitely by storing its earnings for remobilization later, and thereby put off the commitment implied by choosing one of the four. What controls a plant's decision as it chooses among these different possible scenarios? Many of the same things that control our decisions regarding what we do with new resources, it turns out. Our genes limit our possibilities; our environment makes some courses of action wiser than others; some of us are inherently conservative with our earnings; some are prone to gambling; even our fertility status might be considered when evaluating a new plan for investing.

One atmospheric gas in particular—carbon dioxide—is a vital growth resource for plants. Due to the burning of fossil fuels, the level of carbon dioxide within the Earth's atmosphere has increased dramatically in the last fifty-odd years, flooding the plant economy with fast cash and easy credit. Carbon dioxide is the currency of photosynthesis, and plants have now seen decades of an increasingly lavish excess of their most basic resource. We were asking the following question with our radish experiments: What is this going to do to the balance of aboveground versus belowground investment by crops around the world?

Months before, Bill had attached his computer to an inexpensive video camera, and we had been using it to film a set of test plants while they were growing within the chamber. "Check this out," he told me when I arrived in response to his early-morning wake-up call.

The video was a time-lapse compilation of photos taken once every twenty seconds, condensing all of the previous day's growth into four minutes of video. The screen was dark and shadowy at first, indicating that the timed grow lights had not yet switched on. All at once the image lit up and revealed sixteen small potted plants, their stems and leaves limp and relaxed. Shortly into the film, the lights came on and all the plants jarred awake, raising their leaves up toward the light.

One plant located near the edge of the chamber was conspicuous: it twisted and writhed, stretching both upward and outward, shoving the leaves of the adjacent plants out of the way, rudely slapping its broadest leaves down over the central stem of its neighbor. This plant was labeled "C-6" and had started life as a seed of exactly the same size and species as all of the other plants in the chamber. But somehow it *acted* differently from the others while it grew, and at that moment, while watching the video, we were forced to accept what we saw. For several nights now, we had moved C-6 around, changed its neighbors, measured and compared it endlessly, and taken video after video, and the *only* thing different about C-6 was the way it moved after sunrise. While the other plants stretched smoothly and gracefully toward the light, C-6 jerked its smaller leaves feverishly, as if trying to pull itself free of the soil that was holding it.

"I think it hates itself," said Bill.

"I like the little guy. He's got balls," I offered.

"Yeah, well, don't get attached," he advised.

While Bill downloaded and reset the video camera in anticipation of another experiment, I rewatched the video seven or eight times, unable to resist the "smackdown" at about two minutes in, for which we had started to cheer.

"I think he does a little fist-pump right afterward," I observed.

"You're nuts," Bill agreed.

We heard the grow lights switch on behind us, signaling a new day within the chambers, and a vision of the untended paperwork festering on my desk appeared before me.

"Damn it, we'll break him," I decided. "No water for C-6, turn up the lights—and put him in the middle, next to that really big one. Keep the video going."

"Of course," agreed Bill, "it's really the only humane thing to do."

By then the students and postdocs had filtered in, making the whole place confused and busy. From the room behind us we heard a loud clatter and someone hissed, "Ohhh crap," while Bill and I exchanged wry smiles.

"This lab is a well-oiled machine," I announced. "You may as well take your weak ass home and get some sleep."

"Naw," Bill said as he leaned back in his chair. "I want to see how this shit turns out."

C-6 was not part of a formal study, but it changed everything. I had journeyed over some kind of intellectual hill and I could see new territory. We instinctively claimed it using a new language, one that flouted the old rules. Not content with referring to C-6 as "him," we gave him a real name, "Twist and Shout" (which later reverted to "TS-C-6"). We got used to greeting him first thing in the morning and took a kind of sick satisfaction in his ability to endure the torments to which we subjected him. He didn't live for all that long, eventually becoming one of the casualties of Bill's horrible migraine headaches. While Bill was curled up under his desk in the fetal position for ten hours holding his aching head, nothing got watered, fertilized, or filmed, and I tossed C-6 unceremoniously into a waste bin.

Our fascination with C-6 was not a scientifically legitimate experiment, we never officially "wrote it up" for anything, and yet that small plant growing in a Dixie Cup changed my thinking more than anything I had read within my dog-eared textbooks. I had to conclude that C-6 *did* things—not just because he was programmed to do so, but also for reasons known only to him. He could move his "arm" from one side of his "body" to the other; he just did it about 22,000 times slower than I could move mine. His clock and my own were forever out of sync, a simple fact that had placed an untraversable canyon between us. While it seemed that I experienced everything, he appeared to me to passively do nothing. Perhaps, however, to him I was just buzzing around as a blur and, like the electron within an atom, exhibited too much random motion to register as alive.

I stood back and smiled at Bill and all the silly undergrads, and felt the joy that accompanies a new thought as my mind picked up speed like a commuter finally passing a traffic bottleneck. My own spirit had been fed, and at the very least, the day's work ahead of me would be happier because of it. Perhaps that was enough of a scientific accomplishment in itself.

A few hours later I convinced Bill to break for lunch. I told him it would be my treat but that I also had to stop by Whole Foods on an errand. "Me too," he answered, and then explained, "I'm looking into homeopathic remedies for my hand."

We got in my car and drove across the island. Having never actually been inside of a Whole Foods, Bill was immediately enthralled after we walked through the door. He went directly over to a plastic package that cost about thirteen dollars and contained six capers, each the size of a golf ball. He held it up toward me and asked, "Do rich people really eat these things?"

"Absolutely," I answered without looking at what he was showing me. "They love nothing better."

I was occupied in poring over the seven different types of wheatgrass extract available. When I finally identified and selected the greenest one, I noticed that Bill had wandered off, but not before placing the capers inside my cart. I found him marveling at a refrigerated trough of soft French cheese and all at once a plan presented itself. "Let's get all this stuff," I suggested. "Why the hell not?"

"You serious?" Bill had narrowed his eyes dubiously, but his body was tensely hopeful.

"Sure," I announced. "Today we will eat like people with mutual funds."

I often feel guilty that I make more money than Bill does, because our work feels like two halves of one thing. I also like to randomly buy things, and when he's around I can rationalize it as munificence instead of impulsiveness.

"Thank God they had all that crap next to the checkout," Bill observed while reading the label of an organic chocolate bar containing cold-pressed cocoa from the Dominican Republic and açaí berries. "I shudder to think how close I came to missing this thing," he said with his mouth full.

Bill loaded our two-hundred-dollar lunch into my car by himself, shooing away any help from me. He had plans for the four "real thick paper" grocery bags and had begun to hover guardedly over them. He got in on the passenger side and as I started the engine he mumbled, "I sure hope this shit is fair trade," while preparing to unwrap a second chocolate bar, this one flavored with rambutan.

Two hours later we were sitting in the lab and eating "Rockefeller Hot Pockets," which are composed of a slice of *jamón ibérico* wrapped around a spoonful of sturgeon caviar and microwaved for ten seconds. "Crap," I said, startled from checking my watch, "I gotta go, but I'll be back tonight."

Bill waved goodbye with a wedge of Camembert. "See you later." His words were muffled by the baguette that was stuffed in his mouth.

I jumped into my car and raced over to pick my son up from his school, which was just letting out for the day. I traded him his swimsuit and towel for his backpack, and we drove directly to the beach, as was our habit. On the way, I asked him how the third grade was going and he shrugged. We parked in our usual spot across from Kapiolani Park.

Walking across the park, we passed by clusters of great banyan trees, and I stood and waited while he swung from what look like vines but are actually the unanchored roots that grow streaming out of the branches. When we got to the beach, we laid our towels over our shoes and went straight into the ocean and played monk seals for a while, diving and rolling around in the shallows together.

Afterward we sat on the sand and I checked myself for bruises. "Baby monk seals are more rambunctious than the storybooks suggest," I mused while massaging my middle-aged neck. "It's strange that such good swimmers feel the need to ride their parents for locomotion."

My son was digging in the sand. "Are there really animals in there so small that you can't see them?" he asked, referring to the handfuls of wet sand that he was throwing back into the shallows.

"Absolutely," I affirmed. "Tiny animals are everywhere."

"How many?" he asked skeptically.

"Lots," I specified. "Too many to ever count."

He thought for a while, and then said, "I told my teacher that the tiny animals find each other with magnets that are inside their bodies and she said she didn't think so."

I immediately overreacted and retorted defensively, "Well, she's wrong. I know the person who discovered it." I was getting myself worked up.

Like a judge trying to preempt an annoying trial lawyer, he changed the subject. "Well, it doesn't matter anyway because I am going to be a major-league baseball player."

"I promise to come to every single one of your games." I asked my usual question: "Can you get me free tickets?"

He paused for a while, thinking. "Some of them," he finally agreed.

It was getting toward six o'clock, so I stood up, shook out the towels, and gathered up our things, getting ready to leave.

"What's for dessert tonight?" he asked me.

"Your Halloween candy," I replied, and added, "Duh."

He smiled and punched me in the arm.

We went home and I made dinner while he wrestled with our dog, Coco, who is Reba's successor and who, like her, is a Chesapeake Bay retriever. Reba lived to almost fifteen years and was greatly mourned, but through Coco I have come to learn that the entire breed shares her best qualities.

Industrious and indestructible, Coco never hesitates to go out into the rain and is constantly trying to figure out a way to be helpful toward whatever we are doing. She prefers lying on hard cement to lying in her bed, and she will go out back and munch on driveway gravel if she gets hungry before we remember to feed her. She will also run and hurl herself into a seven-foot-high crashing ocean wave if I throw a coconut beyond it and then command her to retrieve, which is what our family does on the weekends. When we travel, she goes and stays at Uncle Bill's house, where she deals severely with the rats that threaten his favorite mango tree.

Clint came home from work just in time for us to eat dinner all together, and afterward we took Coco for a long walk around the neighborhood. Our son was successfully in bed at exactly nine o'clock, but not before I handed him a small vial of wheatgrass juice while he prepared to brush his teeth.

"Drink this first," I commanded. "If you dare," I added.

His eyes widened. "You did it!" he said with awe, and then drank it down while wincing over its bitter taste.

For weeks he had been begging me to make a potion that would turn him into a tiger. "Make it in your lab," he had directed me. "Make it out of plants."

As I tucked him into bed, he got that look that kids wear when there is something important that they want to tell you. "Me and Bill are going to put a basement on our tree house," he told me.

"How are you going to do that?" I asked, genuinely interested.

"We're going to design it," he explained. "It will take a lot of designing. We're going to make a mock-up first."

I pushed my luck. "Can I go inside when it's done?"

"No," he refused, and then reconsidered. "Well, maybe after it's not new anymore." After a pause he closed his eyes and asked, "Am I a tiger yet?"

I looked him up and down slowly, and then answered, "No."

"Why not?" he asked.

"Because it takes a long time," I answered.

"Why does it take a long time?" he pursued.

"Why? I don't know," I admitted, then added, "It takes a long time to turn into what you're supposed to be."

He looked at me as if he wanted to ask more questions, but he also understands that pretending that things are true is often more fun than knowing that they are false.

"But it will work for sure, won't it?" he asked.

"It will work," I confirmed. "It worked before."

"On who?" he said, intrigued.

"On a little mammal named *Hadrocodium,*" I explained. "He lived almost two hundred million years ago and he spent most of his time hiding from the dinosaurs, who would step on him if he didn't watch out. Do you remember the magnolia tree in front of the house where we lived when you were little-little?" I asked.

"That tree out front was the great-great-great-great-and-more-grandchild of the first flower, which looked like it. It was just born as a brand-new kind of plant when *Hadrocodium* was running around.

One day he ate some leaves from it, because his mom told him that it would make him as strong as a dinosaur. But it turned him into a tiger instead. It took a hundred and fifty million years, and a lot of trial and error, but she finally did turn into a tiger."

My son perked up. "'She'? You said it was a 'he.' The tiger is a boy."

"Why can't the tiger be a girl?" I asked.

My son explained the obvious. "Because it's not." After a few seconds he added, "Are you going to the lab tonight?"

"Yes, but I'll be back before you wake up," I assured him. "Daddy is just across the hall, and Coco is watching you while you sleep. This house is full of people who love you," I chanted, our customary bedtime mantra.

He turned to the wall, a signal that he's too sleepy for more talking. I went into the kitchen and made two cups of instant coffee. Looking at the clock, I figured I would get into the lab by ten-thirty. When I picked up my phone to text Bill that I was on my way, I saw that there were already two texts from him. The first read "BRING IPECAC" and the second, sent about an hour later, said "AND MORE FOOD."

I brought the second cup of coffee to Clint and said, "I'm nearly off." We both knew that the pages of handwritten equations that he was busy deriving were thoroughly unintelligible to me, and so he laughed when I said, "Hey, let me know if I can help you with that, eh?"

"Actually," he mentioned, "I would like to get your take on a figure that I made today."

"It's great. I love it," I said without looking up from my purse, as I was digging for my keys.

"It's new. You haven't seen it yet," he emphasized.

"Then it's crap. The y-axis is way off," I said, waving one hand.

He laughed again. "It's a map."

I answered, "Then the colors are wrong. Babe, I gotta go botch my own science; no time to ruin yours." I added helplessly, "The Monkey Jungle never sleeps."

"Well, thanks for the consult," he said as I kissed him.

I went back into our son's room to check that he was sleeping. I kissed him on the forehead and smiled because he had already gotten to the age where he doesn't always let me kiss him when he is awake. I recited the Lord's Prayer and my heart felt full. I petted Coco, who was lying at the foot of the bed; when I hugged her head and whispered, "Will you guard my baby?" she looked at me with the big, somber eyes of a Chesapeake who had answered that question once and for all years ago.

I kissed my husband again, put on my backpack, and went outside to open the shed. I got out my bike and looked up through the warm, tropical sky, into the terminal coldness of space, and saw light that had been emitted years ago from unimaginably hot fires that were still burning on the other side of the galaxy. I put on my helmet and rode to the lab, ready to spend the rest of the night using the other half of my heart.

13

OFTEN WHEN DEALING WITH PLANTS, it is difficult to tell the end from the beginning. Rip almost any plant in half, and its roots can live on for years. The trunk of a felled tree will attempt to grow whole again year after year after year; its inner trunk is lined with dormant buds—sometimes twice as many buds as are visible from the outside—ready to incite. Buds burst as stems, stems become twigs, lucky twigs become branches, good branches persist for decades, and eventually the canopy is as green as it ever was, perhaps all the more so because someone tried to cut it down.

Unlike animals, which function as a single whole, plants are modular in construction, the whole strictly equivalent to the sum of its parts. A tree can shed and replace whole portions of itself and is indeed compelled to do so repeatedly throughout the several centuries of its average life span. In the end, trees die because being alive has simply become too expensive for them. Whenever the sun is up, leaves are working to split water, add atmosphere, and then glue the whole mess into sugar that can be transported down into the stem, where it meets dilute nutrients that were laboriously pulled up by the roots. A plant can bundle all these treasures into new wood and use it to strengthen the trunk or branches.

But the tree also has many other demands: replacing old leaves, making medicine against infection, pumping out flowers and seeds—these use the same raw materials, there are never enough to spare, and there is only so far out or down the tree can go in order to search for them. Eventually it will require more nutrients to maintain the branches and roots that do not grow quite far out enough to capture those nutrients. Once it exceeds the limitations of its environment, it loses all. And this is why you must trim a tree periodically in order to preserve it. Because—as Marge Piercy first said—both life and love are like butter and do not keep: they both have to be made fresh every day.

14

THERE IS SOMETHING PROFOUNDLY SAD about the end of a plant growth experiment. We grow a lot of *Arabidopsis thaliana,* which is a modest little plant. Once it is fully grown you can pick the whole thing up as a single handful. It's one of the very few plants for which scientists have decoded the entire genome, which means that if you unravel the DNA inside one cell of the plant and stretch it out, we can tell you the exact chemical formula of the 125 million molecules that, one after the other, make up the chain.

Once unraveled from its tight snarl within a cell, this chain of proteins stretches almost two full inches. Every single cell in the plant has at least one snarl of these proteins, and scientists have worked out the chemical formula for the whole damn thing. I don't like to think about it, actually; it's just too much data. It overwhelms me. A scientist is supposed to feel overwhelmed at the beginning of her career, not the end. But the more I know, the more my legs buckle underneath me with the weight of all this information.

For the first time in my life, I feel tired. I remember fondly the long weekends of years past when I could work steadily for forty-eight hours, when each new data point reinvigorated me and recharged my mind in stochastic bursts that culminated periodically into new ideas. I still generate ideas, but they are richer and deeper and they come to me while I am sitting down. Such ideas are also much more likely to actually work. And so each morning, I pick up something green and look at it, and then I plant some more seeds. I do it because it is what I know how to do.

Last spring, Bill and I were sifting through the aftermath of a big agricultural experiment up at the greenhouse. We had been growing sweet potatoes under the greenhouse gas levels predicted for the next several hundred years, the levels that we're likely to see if we, as a society, do nothing about carbon emissions. The potatoes grew bigger as carbon dioxide increased. This was not a surprise. We also saw that these big potatoes were less nutritious, much lower in protein content, no matter how much fertilizer we gave them. This was a bit of a surprise. It is also bad news, because the poorest and hungriest nations of the world rely on sweet potatoes for a significant amount of dietary protein. It looks as if the bigger potatoes of the future might feed more people while nourishing them less. I don't have an answer for that one.

The harvest had taken place a few days before with a huge team of students working for almost three days straight, all led by a fantastically strong and wise young man named Matt who would graduate soon. During the course of the experiment he had grown as well, coming into himself as a leader and an expert in a way that was beautiful to see. He could now stand up in front of twenty people in a scene of chaos, streamline each person toward one useful activity, and then provide nonstop advice and quality control for days. It was as if he had gone to war on those plants and the odd leaf or root lying about the place was evidence of his victory. Bill and I had felt truly privileged to stand by, hands-off as we must eventually be when a student nears graduation.

But now it was all over, and everyone was home resting—except us. This is what it must feel like to visit your son's room after he leaves for college: the beginnings of his life left haphazardly behind, irrelevant to him but still precious to you. The air of the greenhouse was thick with the smell of potting soil; Matt had unearthed every potato from every plant and photographed, measured, and described them each individually. The whole thing was a bit of a blur in the harsh light of day; I sensed that I needed to go home and get some rest, but then again, I supposed that a few more hours wouldn't kill me, and so I stayed.

My phone buzzed and I looked at the calendar, only to realize that I was about to miss the mammogram that was three years overdue, which I had already rescheduled once that semester. *Oh crap,* I thought. *Not again.*

The greenhouse door swung open and Bill came in.

"We can cut out our own tumors, right?" I asked him. "I mean, we've got a box cutter around here somewhere, don't we?"

Bill answered without missing a beat, "A drill would work better." He reflected for a moment. "I think I've got a special tip for that, actually."

He was chewing hard on one end of a slice of cold, dry pizza that he'd found in one of the many boxes that had been ordered and discarded during the night. Twenty years, I thought, and Bill wasn't any the worse for wear.

Bill was thinking of something different. He looked at me and asked, "Good Lord, did you age five years while I was outside?" and then added, "You look like a fucking sea hag."

"You're fired," I told him. "Go see the other sea hags down in HR for your paperwork."

"They don't work on Saturdays. Besides, c'mon, you gotta come outside." He motioned toward the door.

The greenhouse that we use is one of many at the university research station, nestled up in the valley beside a little creek that flows down to the ocean. Each greenhouse is as big as a gymnasium and is composed of little more than a huge stainless-steel scaffolding covered over with sheer shade cloth. The Hawaiian Islands are themselves more or less a string of greenhouses: plant growth conditions are excellent year-round, complete with daily showers that are more like routine watering events than like storms.

I looked where Bill was pointing, up toward the jungled mountains, and saw a bright ribbon of rainbow stretching in a full arc across the sky. Its sharp focus made it all the more brazen and beautiful, and it was bracketed by a second rainbow, wider and fuzzier, a gentle halo supporting the confident blaze of the first.

"Hey, it's a double rainbow," I marveled.

"Goddamn right it's a double rainbow," Bill said.

"Well, you don't see them often," I said, justifying my wonder.

"Nope," Bill agreed. "Nobody sees the second rainbow. But it's always there; it's just that nobody sees it. The big rainbow probably thinks that it's alone."

I looked at him hard. "You're certainly deep today," I remarked, and then played my part. "The two rainbows are actually one. A single ray of light moving through bad weather just gives the appearance of two separate things."

Bill paused, and then commented briskly, "Well, rainbows are self-centered fuckers who need to get over themselves."

I observed that that wasn't likely to happen anytime soon.

We walked around back, got a couple of lawn chairs from the old shed, and went back inside the greenhouse. The far side of the huge space was a shambles, with stacks of dirty flowerpots in the corner, one of which had been used as a bucket to hold a big snarl of dirty measuring tape. There was a loose heap of soil in one spot, and we set up our folding chairs next to it and then sat with our bare feet in the cool, damp dirt. At the other end of the greenhouse was somebody else's ongoing experiment. Perennial in every sense, it had been there before we came and will probably still be going when I retire.

"How can you not like that?" I waved an arm toward the rows and rows of profuse orchids. "Just smell it."

"We've got it pretty damn good, I have to admit," said Bill. "Never dreamed I'd end up in Hawaii," he continued.

I worry about Bill. I worry about his past, and his would-haves. I worry that he would have a wife and a bunch of kids if he hadn't been hanging around me all these years. Bill keeps explaining to me that because Armenians commonly live for more than a hundred years and he's not even fifty yet, he's still too young to start dating. Nevertheless, I worry about his future. I worry that when he does meet someone, she won't be good enough for him. Bill always laughs this off. "Women used to be put off that I lived in a van," he complains; "now they only want me for my money."

Bill is indeed living well. His house rests high on a hill overlooking Honolulu; his homegrown mangoes are the jewel in the crown of his rich and ever-blooming garden. Bill accidentally made a small fortune when he sold the Baltimore house that he had bought as a monstrosity with rotting pipes, shoddy electrical, and a melted foundation—all of which he fixed, late at night and without help, turning it into a gorgeous piece of university-adjacent real estate.

People still puzzle over the two of us, Bill and me. Are we siblings? Soul mates? Comrades? Novitiates? Accomplices? We eat almost every meal together, our finances are mixed, and we tell each other everything. We travel together, work together, finish each other's sentences, and have risked our lives for each other. I'm happily married with a family and Bill was an obvious precondition to all that, a brother whom I would never give up, part of the package. But people that I meet still seem to want a label for what is between us. Just as with the potatoes, I don't have an answer for that one. I do us because us is what I know how to do.

I reached over and picked up a watering can, raining water over the soil covering both our feet. We wiggled our toes and worked the dirt into a nice luxurious mud, and then we leaned back and just sat for a while. Bill eventually broke the silence with "So! What should we do now? We're good until 2016, right?"

Bill was referring to our funding for the lab; we were indeed financially solid through the summer of 2016, with several federal government contracts in place. After that, however, the lab could still fold: research funding for environmental science decreases every year. I have tenure, but Bill certainly doesn't—that sort of thing is only for professors. It is maddening to me that the best and hardest-working scientist I've ever known has no long-term job security, and that this is mostly my fault. The only thing that I can think to do if I lose funding is to threaten to quit, which would probably just leave both of us out on the street. As research scientists, we will never, ever be secure.

"Hey, snap out of it." Bill clapped his hands in front of my face. "What should we do next? We can do anything we want!" He rubbed his hands together, slapped his thighs, and stood up. Bill was right, as usual. *O me of little faith.* What hard-working team anywhere doing anything has any more security than we do? We will be like the lilies of the field, I decided, except that we *will* toil and spin and sow and reap.

I stood up and stepped forward. "Well, what have we got?" I glanced around, taking a casual inventory of our scattered equipment. "I know," I said, "let's put all of our stuff in a big stockpile and stare at it for a while. Something will come to me."

Bill nodded at me and walked to the other side of the greenhouse. He brought over the stash of grow lights that were still good and put them down gently beside the wads of extension cords that I had dragged over from the other side. Then we worked together to move the miter saw, as well as several uncut two-by-fours and a barrel of particleboard scraps. I brought over our toolboxes and positioned them prominently, one with the lid propped open like a deep-sea treasure chest. Bill slid over a few bags of potting soil and set a bag of fertilizer next to each one.

I was laying out the different seeds that we had, one package next to the other, when I looked up to find Bill dragging toward me a roll of chicken wire that had probably been rusting in the corner for years. I wrinkled my nose. "That's not even ours," I said with disgust.

"It is now," said Bill, and then we both knew what was coming. We began to sneak through the orchid experiment, plucking loose hoses and broken clamps, shoving them into the makeshift aprons of our T-shirts and walking them back to the pile.

"Holy shit," exclaimed Bill as he spied an expensive cordless power drill set down between two orchid plants. Bill and I locked eyes as he picked it up. We have at least five cordless drills already and Bill knows that we could buy any number of them just for the hell of it. We very likely have several times more grant money than whoever owns this tool. Every moral and rational fact argued that we should not have stolen the drill. Except for one: whoever owned it was not there.

"Well, you know what they say about Hell," I remarked while adding the drill to our pile. "The ambiance is bad, but the company is actually pretty good." Bill sat back down and cracked open a Pepsi. I circled the pile, tucking orchid flowers into it here and there as if I were decorating a Christmas tree.

The drill turned out to be broken: it didn't work then and we've never been able to fix it. But we still have it in the lab somewhere—Bill and I have never even considered putting it back or throwing it away. I'll never concede that any tool is useless and I'll never admit that there is one that I don't need. I will never stop being ravenously hungry for science, no matter how well it feeds me.

On that day that Bill and I sat together inside the greenhouse, we began to talk about our hopes and goals, about what plants can do and about what we might be able to make them do. Soon our brainstorming about what to do next included inevitable discussions of what we'd done before. Before long we were telling each other the stories of this book. I am amazed to realize that these stories now span about twenty years.

During that time we've gotten three degrees, worked six jobs, lived in four countries and traveled through sixteen more, ended up in the hospital five times, owned eight old cars, driven at least twenty-five thousand miles, put a dog to sleep, and produced roughly sixty-five thousand carbon stable isotope measurements. This last was our ostensible goal throughout it all. Before we made said measurements, only God and the Devil himself knew what the values were, and we suspect that neither one of them much cared. Now anyone with a library card can look these values up, because we published them as seventy separate articles within forty different journals. We think of this as progress because it is our impossible job to manufacture new information out of whole cloth. Along the way, we also managed to become adults without ceasing to be children. Nothing reminds us of this as well as the stories that we told and retold on that day.

At the end of a long silence Bill surprised me by saying with quiet seriousness, "Put it in a book. Do me that favor someday."

Bill knows about my writing. He knows about the pages of poetry stuffed into my car's glove box; he knows about the many nextstory.doc files on my hard drive; he knows how I like to sift through the thesaurus for hours; he knows that nothing feels better to me than finding exactly the right word that stabs cleanly at the heart of what you are trying to say. He knows that I read most books twice or more and write long letters to their authors, and that sometimes I even get an answer. He knows how much I need to write. But he had never given me permission to write about us until that day. I nodded and inwardly vowed to do my best.

I'm good at science because I'm not good at listening. I have been told that I am intelligent, and I have been told that I am simple-minded. I have been told that I am trying to do too much, and I have been told that what I have done amounts to very little. I have been told that I can't do what I want to do because I am a woman, and I have been told that I have only been allowed to do what I have done because I am a woman. I have been told that I can have eternal life, and I have been told that I will burn myself out into an early death. I have been admonished for being too feminine and I have been distrusted for being too masculine. I have been warned that I am far too sensitive and I have been accused of being heartlessly callous. But I was told all of these things by people who can't understand the present or see the future any better than I can. Such recurrent pronouncements have forced me to accept that because I am a female scientist, nobody knows *what* the hell I am, and it has given me the delicious freedom to make it up as I go along. I don't take advice from my colleagues, and I try not to give it. When I am pressed, I resort to these two sentences: You shouldn't take this job too seriously. Except for when you should.

I have accepted that I don't know all the things that I ought to know, but I do know the things that I need to know. I don't know how to say "I love you," but I do know how to show it. The people who love me know the same.

Science is work, nothing more and nothing less. And so we will keep working as another day dawns and this week turns into next week, and then this month becomes next month. I can feel the warmth of the same brilliant sun that shines above the forest and onto the green world, but in my heart I know that I am not a plant. I am more like an ant, driven to find and carry single dead needles, one after the other, all the way across the forest and then add them one by one to a pile so massive that I can only fully imagine one small corner of it.

As a scientist I am indeed only an ant, insufficient and anonymous, but I am stronger than I look and part of something that is much bigger than I am. Together we are building something that will fill our grandchildren's grandchildren with awe, and while building we consult daily the crude instructions provided by our grandfathers' grandfathers. As a tiny, living part of the scientific collective, I've sat alone countless nights in the dark, burning my metal candle and watching a foreign world with an aching heart. Like anyone else who harbors precious secrets wrought from years of searching, I have longed for someone to tell.

Epilogue

PLANTS ARE NOT LIKE US. They are different in critical and fundamental ways. As I catalog the differences between plants and animals, the horizon stretches out before me faster than I can travel and forces me to acknowledge that perhaps I was destined to study plants for decades only in order to more fully appreciate that they are beings we can never truly understand. Only when we begin to grasp this deep otherness can we be sure we are no longer projecting ourselves onto plants. Finally we can begin to recognize what is actually happening.

Our world is falling apart quietly. Human civilization has reduced the plant, a four-hundred-million-year-old life form, into three things: food, medicine, and wood. In our relentless and ever-intensifying obsession with obtaining a higher volume, potency, and variety of these three things, we have devastated plant ecology to an extent that millions of years of natural disaster could not. Roads have grown like a manic fungus, and the endless miles of ditches that bracket these roads serve as hasty graves for perhaps millions of plant species extinguished in the name of progress. Planet Earth is nearly a Dr. Seuss book made real: every year since 1990 we have created more than eight billion new stumps. If we continue to fell healthy trees at this rate, less than six hundred years from now, every tree on the planet will have been reduced to a stump. My job is about making sure there will be some evidence that someone cared about the great tragedy that unfolded during our age.

In languages across the globe, the adjective "green" is etymologically rooted in the verb "to grow." In free-association studies, participants linked the word "green" to concepts of nature, restfulness, peace, and positivity. Research has shown how a brief glimpse of green significantly improved the creativity that people brought to bear on simple tasks. Viewed from space, our planet appears less green with each passing year. On my bad days, our global troubles seem only to have increased over my lifetime, and I can't escape my greatest nagging fear: When we are gone, will we leave our heirs stranded in a pile of rubble, just as sick and hungry and war-exhausted as we ever were, bereft even of the homely comfort of the color green? But on my good days, I feel like I can do something about this.

Every single year, at least one tree is cut down in your name. Here's my personal request to you: If you own any private land at all, plant one tree on it this year. If you are renting a place with a yard, plant a tree in it and see if your landlord notices. If he does, insist to him that it was always there. Throw in a bit about how exceptional he is for caring enough about the environment to have put it there. If he takes the bait, go plant another one. Baffle some chicken wire at its base and string a cheesy birdhouse around its tiny trunk to make it look permanent, then move out and hope for the best.

There are more than one thousand successful tree species for you to choose from, and that's just for North America. You will be tempted to choose a fruit tree because they grow quickly and make beautiful flowers, but these species will break under moderate wind, even as adults. Unscrupulous tree planting services will pressure you to buy a Bradford pear or two because they establish and flourish in one year; you'll be happy with the result long enough for them to cash your check. Unfortunately, these trees are also notoriously weak in the crotch and will crack in half during the first big storm. You must choose with a clear head and open eyes. You are marrying this tree: choose a partner, not an ornament.

How about an oak? There are more than two hundred species and one is bound to be adapted to your specific corner of the planet. In New England, the pin oak thrives, its leaves tipping to a thorny point in a good-natured impression of its evergreen neighbor the holly bush. The turkey oak can grow practically submerged within the wetlands of Mississippi, its leaves soft as a newborn's skin. The live oak can grow sturdily on the hottest hills of central California, contrasting dark green against the golden grass. For my money, I'll take the bur oak, the slowest-growing but the strongest of all; even its acorns are heavily armored, ready to do battle with the uninviting soil.

Speaking of money, you may not even need any: Several state and local agencies have embarked upon tree-planting programs, distributing seedlings for free or at a reduced cost. For example, the New York Restoration Project provides trees as part of its goal to help citizens plant and care for one million new trees across New York City's five boroughs, while the Colorado State Forest Service provides access to its nurseries to any local landowner holding one or more acres. Every state university runs one or more large operations called Extension Units, full of experts qualified to give advice and encouragement to citizen gardeners, tree owners, and nature enthusiasts of all types. Call around: these researchers are obligated to provide free consultations to interested civilians regarding your trees, your compost heap, your out-of-control poison ivy.

Once your baby tree is in the ground, check it daily, because the first three years are critical. Remember that you are your tree's only friend in a hostile world. If you do own the land that it is planted on, create a savings account and put five dollars in it every month, so that when your tree gets sick between ages twenty and thirty (and it will), you can have a tree doctor over to cure it, instead of just cutting it down. Each time you blow the account on tree surgery, put your head down and start over, knowing that your tree is doing the same. The first ten years will be the most dynamic of your tree's life; what kind of overlap will it make with your own? Take your children to the tree every six months and cut a horizontal chink into the bark to mark their height. Once your little ones have grown up and moved out and into the world, taking parts of your heart with them, you will have this tree as a living reminder of how they grew, a sympathetic being who has also been deeply marked by their long, rich passage through childhood.

While you're at it, would you carve Bill's name into your tree as well? He's told me a hundred times over that he'll never read this book because it would be pointless. He says that if he ever gets at all interested in himself he can damn well sit down and remember the last twenty years without any help from me. I don't have a good comeback for that one, but I'd like to think that the many parts of Bill that I've released to the wind belong somewhere, and over the years we've learned that the best way to give something a home is to make it part of a tree. My name is carved into a bunch of our lab equipment, so why shouldn't Bill's name be carved into a bunch of trees?

At the end of this exercise, you'll have a tree and it will have you. You can measure it monthly and chart your own growth curve. Every day, you can look at your tree, watch what it does, and try to see the world from its perspective. Stretch your imagination until it hurts: What is your tree trying to do? What does it wish for? What does it care about? Make a guess. Say it out loud. Tell your friend about your tree; tell your neighbor. Wonder if you are right. Go back the next day and reconsider. Take a photograph. Count the leaves. Guess again. Say it out loud. Write it down. Tell the guy at the coffee shop; tell your boss.

Go back the next day, and the next, and so on. Keep talking about it; keep sharing its unfolding story. Once people begin to roll their eyes and gently tell you that you're crazy, laugh with gratification. When you're a scientist, it means that you're doing it right.

Acknowledgments

Writing *Lab Girl* has been the most joyful work of my life, and I am grateful to those who helped and supported me. I thank everyone at Knopf, especially my editor, Robin Desser. This is a better book and I am a better writer because of the care she has taken. Tina Bennett has been more than my agent: she taught me the difference between a bunch of stories and a book. My great debt to her is my most precious professional possession. Svetlana Katz was my lifeline for years while I was searching for the style of this narrative. She never doubted and so I kept the faith. No words can describe the gratitude that a hopeful writer feels for the first known author who reads her work and then encourages her. For me, that person was Adrian Nicole LeBlanc. I can name no deeper comfort than the friendship of those who knew me as a child. Thank you, Connie Luhmann, for being my eyes when I needed you. I am also grateful to Heather Schmidt, Dan Shore, and Andy Elby, who after reading some, always came back and asked to read more.

Endnote

Every book about plants is a story without an ending. For each of the facts that I've shared with you, there are at least two baffling mysteries that I'm aching to solve. Can grown trees recognize their own seedlings? Is there plant life on other planets? Did the very first flowers make the dinosaurs sneeze? All of those questions will have to wait for another day. But here I can't resist adding a few more details about how I figured out some of the content and presented it.

A good deal of the information about plants in *Lab Girl* was derived from calculations that I acquired the habit of making during my twenty-plus years of teaching, in order to help facts "stick" in the minds of my students. For example, this sentence in chapter 9 of Part Two: "In the United States alone, the total length of the wooden planks used during the last twenty years was more than enough to build a footbridge from the planet Earth to the planet Mars" (page 77) was drawn from a simple comparison of lumber consumption statistics as reported by the U.S. Department of Commerce (805 billion board feet used between 1995 and 2010), with the average distance from Earth to the planet Mars as reported by NASA (140 million miles, which equals 739 billion feet). Other places where I have accessed similar facts or statistics for this book include the U.S. Census Bureau, the U.S. Forest Service, the U.S. Department of Agriculture, the National Center for Health Statistics, and the Food and Agriculture Organization of the United Nations.

Making certain of the calculations in *Lab Girl* was of course complicated by the fact that every conceivable attribute one can measure for a particular plant reveals a vast variation, when compared to other plants of different species. To illustrate: in order to make the calculation presented in chapter 3 of Part One about the relative abundance of growing plants versus waiting seeds, I pictured myself in a deciduous forest, and so estimated 500 seeds lying dormant in the soil beneath each of my footsteps. Had I instead chosen to picture myself walking through a grassland, I would have estimated more than 5,000 seeds beneath each footstep, due to the fact that grass seeds are much, much smaller than those dispersed by trees—a big difference. So while writing *Lab Girl,* I held myself to the following policy: whenever presented with such a choice, I picked the scenario where the scope of variation yielded a more modest result. Therefore, I would ask a reader to bear in mind that each of my claims about plants, impressive and marvelous as some of them may appear, were set up to "err" on the side of understatement.

My calculations regarding a "modest, unremarkable tree" described in chapter 5 of Part Two were based upon a real tree, familiar and dear to my heart: a small candlenut *(Aleurites moluccanus),* very similar in appearance and function to the more common maple. The little candlenut is one of the trees growing in the courtyard outside my laboratory at the University of Hawaii. For many years I taught a class called Terrestrial Geobiology, and at the end of each lecture, the students and I would go outside to visit the tree and reflect upon it as an illustration of the day's material. As one of the homework exercises for the course, the students and I measured the various properties (total height, leaf density, carbon content, etc.) that allowed us to calculate how much water, sugar, and nutrients the tree requires every growing season—that is, the information that I've presented on pages 120–21.

In my description of federal funding in the United States for "curiosity-driven research" found within chapter 5 of Part Two (pages 121–25), I used data from Fiscal Year 2013, because it seemed to best reflect the most recent and complete datasets across multiple government agencies. However, it matters little which year I used for my analysis, as the total federal allocation to the National Science Foundation has not meaningfully increased for more than a decade. Similarly, my statement on page 123 that "the amount of the U.S. annual budget that goes to non-defense-related research has been frozen" is based on data compiled by the American Association for the Advancement of Science, which revealed that for every year since 1983, total spending on scientific research has comprised a flat 3 percent of the total United States federal budget.

In studying plants, I am fortunate to work within a field populated by exceptionally creative and prolific researchers, and I relish the time that I spend reading about studies performed by my peers. I worked the story of my "top three" such studies into the pages of *Lab Girl,* and I want to credit the scientists behind the original experiments:

The Sitka willow experiments described on pages 166–67 were first published by D. F. Rhoades in 1983. It was not until 2004, more than twenty years later, that G. Arimura and coauthors showed how VOC production in one plant could affect gene expression within a separate plant upon exposure, and thus demonstrated the mechanism by which the willow trees communicated with one another.

The phenomenon that scientists call "hydraulic lift"—or water moving "up from the strong and out toward the weak" as I describe it on page 231—was first shown by Dawson (1993) within sugar maple *(Acer saccharum)*.

It was Kvaalen and Johnsen (2008) who demonstrated that *Picea abies* "remembered their cold seedhoods"—as I put it on page 232—by comparing juvenile trees that had been cultured as embryos under different temperatures and then grown for years within the same greenhouse.

And finally, for readers who find themselves wanting to know more about the living green that surrounds us, I recommend that they waste no time in getting ahold of P. A. Thomas's book *Trees: Their Natural History* (2000), a clearly written introductory textbook full of fascinating information. Whenever people tell me that they are interested in learning more about deforestation, or about global change in general, I point them toward the illuminating Vital Signs series, which is the annual publication of the Worldwatch Institute (www.worldwatch.org), a nongovernmental organization and independent research institute founded in 1974 that analyzes the ongoing changes, trends, and global patterns found in the data collected each year by multiple agencies within the U.S. Energy Information Administration, the International Energy Agency, the World Health Organization, the World Bank, the United Nations Development Programme, as well as the Food and Agriculture Organization of the United Nations and many other agencies.

WORKS CITED

Arimura, G., D. P. Huber, and J. Bohlmann. 2004. Forest tent caterpillars *(Malacosoma disstria)* induce local and systemic diurnal emissions of terpenoid volatiles in hybrid poplar *(Populus trichocarpa × deltoides)*: cDNA cloning, functional characterization, and patterns of gene expression of (−)-germacrene D synthase, PtdTPS1. *Plant Journal* 37 (4):603–16.

Dawson, T. E. 1993. Hydraulic lift and water use by plants: Implications for water balance, performance and plant-plant interactions. *Oecologia* 95 (4):565–74.

Kvaalen, H., and Ø. Johnsen. 2008. Timing of bud set in *Picea abies* is regulated by a memory of temperature during zygotic and somatic embryogenesis. *New Phytologist* 177 (1):49–59.

Rhoades, D. F. 1983. Responses of alder and willow to attack by tent caterpillars and webworms: Evidence for pheromonal sensitivity of willows. In *Plant resistance to insects,* ed. P. A. Hedin, 55–68. Washington, D.C.: American Chemical Society.

Thomas, P. A. 2000. *Trees: Their natural history.* Cambridge and New York: Cambridge University Press.